Disclaimer

The publisher of this book is by no way associated with the National Institute of Standards and Technology (NIST). The NIST did not publish this book. It was published by 50 page publications under the public domain license.

50 Page Publications.

Book Title: Economics of Egress Alternatives and Life-Safety Costs

Book Author: Robert E. Chapman; David T. Butry; Allison L. Huang; Douglas S. Thomas;

Book Abstract: Fire protection measures are needed to maintain the safety and integrity of the Nation s building stock and to limit loss of life and property when building fires do occur. Statistics published by the National Fire Protection Association demonstrate that fire protection is a major investment cost in building construction. Therefore, ways to reduce these costs while ensuring safety are of interest to building owners, fire protection engineers, and other construction industry stakeholders. Although all fire protection measures have important economic implications, the focus of this report is on egress-related requirements in new building construction. Recent changes in the International Building Code have set the stage for analyzing the costs of several key egress-related requirements. The U.S. General Services Administration commissioned this study to conduct an economic analysis of the use of elevators and exit stairs for occupant evacuation and fire service access. The goal of this study is to produce analyses of cost data suitable for evaluating improved egress system designs that promote efficient and timely egress of occupants, including those with disabilities, and that facilitate more efficient fire department operations. This report tabulates cost data for selected egress-related requirements in five prototypical buildings. The five prototypical buildings range in height from a 5-floor, mid-rise building to a 75-floor, high-rise building. Cost data are tabulated in a format that facilitates life-cycle cost analyses of selected egress-related requirements. Incremental costs are also tabulated to help assess the implications of changing one or more design parameters.

Citation: NIST SP - 1109

Keyword: Buildings; cost data; economic analysis; egress; exit stairs; fire protection; life-cycle cost; occupant evacuation elevators; safety

NIST Special Publication 1109

U.S. Department of Commerce
National Institute of Standards and Technology

Office of Applied Economics
Building and Fire Research Laboratory
Gaithersburg, Maryland 20899

Economics of Egress Alternatives and Life-Safety Costs

Robert E. Chapman, David T. Butry, Allison L. Huang, and Douglas S. Thomas

NIST Special Publication 1109

U.S. Department of Commerce
National Institute of Standards and Technology

Office of Applied Economics
Building and Fire Research Laboratory
Gaithersburg, Maryland 20899-8603

Economics of Egress Alternatives and Life-Safety Costs

Robert E. Chapman, David T. Butry, Allison L. Huang, and Douglas S. Thomas

Sponsored by:

U.S. General Services Administration
Public Buildings Service
Office of Facilities Management and Services Programs
Washington, DC 20405

September 2010

U.S. DEPARTMENT OF COMMERCE
Gary Locke, Secretary

NATIONAL INSTITUTE OF STANDARDS AND TECHNOLOGY
Patrick D. Gallagher, Director

Abstract

Fire protection measures are needed to maintain the safety and integrity of the Nation's building stock and to limit loss of life and property when building fires do occur. Statistics published by the National Fire Protection Association demonstrate that fire protection is a major investment cost in building construction. Therefore, ways to reduce these costs while ensuring safety are of interest to building owners, fire protection engineers, and other construction industry stakeholders. Although all fire protection measures have important economic implications, the focus of this report is on egress-related requirements in new building construction.

Recent changes in the International Building Code have set the stage for analyzing the costs of several key egress-related requirements. The U.S. General Services Administration commissioned this study to conduct an economic analysis of the use of elevators and exit stairs for occupant evacuation and fire service access. The goal of this study is to produce analyses of cost data suitable for evaluating improved egress system designs that promote efficient and timely egress of occupants, including those with disabilities, and that facilitate more efficient fire department operations. This report tabulates cost data for selected egress-related requirements in five prototypical buildings. The five prototypical buildings range in height from a 5-floor, mid-rise building to a 75-floor, high-rise building. Cost data are tabulated in a format that facilitates life-cycle cost analyses of selected egress-related requirements. Incremental costs are also tabulated to help assess the implications of changing one or more design parameters.

The results of the economic analysis for four prototypical buildings over 120 ft (37 m), with two over 420 ft (128 m) high, demonstrate that: (1) an additional exit stair is a cost-effective alternative to the installation of occupant evacuation elevators on a first-cost basis; and (2) occupant evacuation elevators are a cost-effective alternative to the installation of an additional exit stair on a life-cycle cost basis when rental rates are high and discount rates are low.

Keywords

Buildings; cost data; economic analysis; egress; exit stairs; fire protection; life-cycle cost; occupant evacuation elevators; safety

Preface

This study was conducted by the Office of Applied Economics in the Building and Fire Research Laboratory at the National Institute of Standards and Technology. This report produces analyses of cost data suitable for evaluating improved egress system designs that promote efficient and timely egress of occupants, including those with disabilities, and that facilitate more efficient fire department operations. The intended audience is the National Institute of Standards and Technology, the U.S. General Services Administration, building owners, fire protection engineers, standards and codes developers, and other construction industry stakeholders interested in reducing the costs of fire protection while ensuring safety.

Disclaimer

Certain trade names and company products are mentioned in the text in order to adequately specify the technical procedures and equipment used. In no case does such identification imply recommendation or endorsement by the National Institute of Standards and Technology, nor does it imply that the products are necessarily the best available for the purpose.

Disclaimer Regarding Non-Metrics Units

The policy of the National Institute of Standards and Technology is to use metric units in all of its published materials. Because this report is intended for the U.S. construction industry that uses U.S. customary units, it is more practical and less confusing to use U.S. customary units rather than metric units. Measurement values in this report are therefore stated in U.S. customary units first, followed by the corresponding values in metric units within parentheses.

Cover Photographs Credits

Digital Division Construction in Action clip gallery image used in compliance with Digital Vision's royalty free digital stock photography use policy.

Acknowledgements

The authors wish to thank all those who contributed so many excellent ideas and suggestions for this report. They include Mr. David W. Frable, Senior Fire Protection Engineer, U.S. General Services Administration (GSA), for his technical guidance, suggestions, and support; Messrs. Jason D. Averill and Richard D. Peacock of the Fire Research Division in the Building and Fire Research Laboratory at the National Institute of Standards and Technology (NIST) for coordinating the GSA/NIST egress research program; and Dr. S. Shyam Sunder, Director of the Building and Fire Research Laboratory (BFRL) at NIST, for his guidance and insights. Special appreciation is extended to Dr. Harold E. Marshall of BFRL's Office of Applied Economics for his thorough review and many insights and to Ms. Carmen L. Pardo for her assistance in preparing the manuscript for review and publication. Special appreciation is also extended to Mr. William H. Hunt, GSA's Chief Estimator; Mr. Ahmet Tanyeri, President Van Dusen and Associates; Mr. Carl Galioto, Partner Skidmore, Owings & Merrill LLP; Ms. Barbara Balboni, Senior Engineer RS Means; Ms. Vivian Huang, Associate Principal Kohn Pedersen Fox Associates; Mr. Muthiah Kasi, CEO Alfred Benesch & Company, and Mr. Brian D. Black, Code and Safety Consultant to National Elevator Industry, Inc., BDBlack Codes, Inc., for their help in developing much of the cost data that is the focus of this report. Special appreciation is also extended to Mr. Richard Bukowski, formerly with BFRL's Fire Research Division, and Dr. John R. Hall, Jr., Division Director for Fire Analysis & Research at the National Fire Protection Association, for their comments on an earlier draft of this report. The report has also benefitted from the review and technical comments provided by Dr. Nicos S. Martys of BFRL's Materials and Construction Research Division and Dr. Anthony P. Hamins, Chief of BFRL's Fire Research Division.

Table of Contents

Abstract iii
Preface v
Acknowledgements vii
Acronyms and Abbreviations xiii
Executive Summary xv
1   Introduction 1
   1.1   Background 1
   1.2   Purpose 3
   1.3   Scope and Approach 3
   1.4   Assumptions and Limitations 4
2   Annual Cost of Fire Protection 5
   2.1   Value of Construction Put in Place 5
   2.2   Characteristics of Commercial Buildings 7
3   Methodology for Conducting an Economic Analysis of Egress- Related Costs. 15
   3.1   Types of Analysis 15
      3.1.1   Baseline Analysis 15
      3.1.2   Sensitivity Analysis 16
   3.2   Overview of Evaluation Methods 16
      3.2.1   Life-Cycle Cost Method 17
      3.2.2   Present Value Net Savings 18
      3.2.3   Savings-to-Investment Ratio 18
      3.2.4   Adjusted Internal Rate of Return 19
   3.3   Presentation and Analysis of the Results of an Economic Analysis 19
      3.3.1   Significance of Study Effort 20
      3.3.2   Analysis Strategy 20
      3.3.3   Calculation of Benefits, Costs, and Additional Measures 23
4   Tabulation and Analysis of Egress-Related Cost Data 25
   4.1   Use of Prototypical Building Designs 25
   4.2   Exit Stairs 25
      4.2.1   Alternative Exit Stair Configurations 26
      4.2.2   Incremental Cost of Adding an Exit Stair 33
   4.3   Occupant Evacuation Elevators and Fire Service Access Elevators 39
      4.3.1   Occupant Evacuation Elevators 40
      4.3.2   Fire Service Access Elevators 44
   4.4   Cost-Effectiveness Analysis: Additional Exit Stair or Occupant Evacuation Elevators 47
      4.4.1   Results of the Baseline Analysis 49
      4.4.2   Results of the Sensitivity Analysis 59
          4.4.2.1   Uncertainty Parameters 60
          4.4.2.2   Monte Carlo Simulation of the Life-Cycle Costs of an Additional Exit Stair and Occupant Evacuation Elevator 61

  4.4.2.3 Monte Carlo Simulation of the Economic Performance of Occupant Evacuation Elevators Compared to an Additional Exit Stair: PVNS, SIR, and AIRR 62
    4.4.2.4 Break-Even Analysis... 69
 4.4.3 Presentation and Analysis of Results.. 70

5 Summary and Recommendations for Further Research 81
5.1 Summary... 81
5.2 Recommendations for Further Research.. 82

References 83

Appendix A Summary of Cost Estimates for Converting Passenger Elevators to Occupant Evacuation Elevators 87

Appendix B Summary of Cost Estimates for Converting Service Elevators to Fire Service Access Elevators 101

List of Figures

Figure 2.1 Number of Commercial Buildings by Size Category: 2003 10
Figure 2.2 Number of Commercial Buildings by Year of Construction: 2003 10
Figure 2.3 Total Floorspace of Commercial Buildings by Size Category: 2003 12
Figure 2.4 Total Floorspace by Size Category and Number of Floors for All Commercial Buildings: 2003 12
Figure 2.5 Breakdown of High-rise Commercial Buildings by Number 13
Figure 2.6 Breakdown of High-rise Commercial Buildings by Floorspace 13
Figure 4.1 Plan and Cross Section of Exit Stair Used in Calculating Lost Rental Income: Exit Stair with Nominal Width of 44 in (112 cm) 34

List of Tables

Table ES.1 Summary Information on the Prototypical Buildings Used in Developing Egress-Related Cost Data xviii
Table ES.2 Assumptions for the Monte Carlo Simulations... xix
Table 2.1 Value of Construction Put in Place for Selected Construction Types: 2002 to 2008.. 6
Table 2.2 Annual Cost of Fire Protection in Buildings: 2002 to 2008 7
Table 2.3 Number of Commercial Buildings and Total Floorspace by Principal Building Activity 8
Table 4.1 Summary Information on the Prototypical Buildings Used in Developing Egress-Related Cost Data 25
Table 4.2 Exit Stair-Related Cost Data for Building 1: 5 Floors, Height of 60 ft (18 m), and Total Floorspace of 100 000 ft^2 (9290 m^2)... 28
Table 4.3 Exit Stair-Related Cost Data for Building 2: 13 Floors, Height of 156 ft (48 m), and Total Floorspace of 325 000 ft^2 (30 193 m^2) 29
Table 4.4 Exit Stair-Related Cost Data for Building 3: 28 Floors, Height of 336 ft (102 m), and Total Floorspace of 840 000 ft^2 (78 038 m^2) 30

Table 4.5 Exit Stair-Related Cost Data for Building 4: 42 Floors, Height of 504 ft (154 m), and Total Floorspace of 1 680 000 ft^2 (156 076 m^2) 31

Table 4.6 Exit Stair-Related Cost Data for Building 5: 75 Floors, Height of 900 ft (274 m), and Total Floorspace of 3 375 000 ft^2 (313 545 m^2) 32

Table 4.7 Incremental Cost of Adding an Exit Stair 33

Table 4.8 Incremental Cost per Floor of Adding an Exit Stair 34

Table 4.9 Loss of Rental Space Due to the Installation of an Additional Exit Stair 36

Table 4.10 Annual Loss of Rental Income Due to the Installation of an Additional Exit Stair 37

Table 4.11 Life-Cycle Costs of an Additional Exit Stair 38

Table 4.12 Incremental Cost of Converting Passenger Elevators to Occupant Evacuation Elevators for Building 2: 13 Floors, Height of 156 ft (48 m), and Total Floorspace of 325 000 ft^2 (30 193 m^2) 41

Table 4.13 Incremental Cost of Converting Passenger Elevators to Occupant Evacuation Elevators for Building 3: 28 Floors, Height of 336 ft (102 m), and Total Floorspace of 840 000 ft^2 (78 038 m^2) 42

Table 4.14 Incremental Cost of Converting Passenger Elevators to Occupant Evacuation Elevators for Building 4: 42 Floors, Height of 504 ft (154 m), and Total Floorspace of 1 680 000 ft^2 (156 076 m^2) 42

Table 4.15 Incremental Cost of Converting Passenger Elevators to Occupant Evacuation Elevators for Building 5: 75 Floors, Height of 900 ft (274 m), and Total Floorspace of 3 375 000 ft^2 (313 545 m^2) 43

Table 4.16 Life-Cycle Costs of Converting a Standard Passenger Elevator System to an Occupant Evacuation Elevator System 44

Table 4.17 Incremental Cost of Converting Service Elevators to Fire Service Access Elevators for Building 2: 13 Floors, Height of 156 ft (48 m), and Total Floorspace of 325 000 ft^2 (30 193 m^2) 45

Table 4.18 Incremental Cost of Converting Service Elevators to Fire Service Access Elevators for Building 3: 28 Floors, Height of 336 ft (102 m), and Total Floorspace of 840 000 ft^2 (78 038 m^2) 45

Table 4.19 Incremental Cost of Converting Service Elevators to Fire Service Access Elevators for Building 4: 42 Floors, Height of 504 ft (154 m), and Total Floorspace of 1 680 000 ft^2 (156 076 m^2) 46

Table 4.20 Incremental Cost of Converting Service Elevators to Fire Service Access Elevators for Building 5: 75 Floors, Height of 900 ft (274 m), and Total Floorspace of 3 375 000 ft^2 (313 545 m^2) 46

Table 4.21 Life-Cycle Costs of Converting a Standard Service Elevator System to a Fire Service Access Elevator System 47

Table 4.22A Summary of Key Life-Cycle Cost Measures for an Additional Exit Stair and Occupant Evacuation Elevators 51

Table 4.22B Summary of Key Life-Cycle Cost Measures for an Additional Exit Stair and Occupant Evacuation Elevators 52

Table 4.23A Calculation of Present Value Net Savings of Occupant Evacuation Elevators vis-à-vis an Additional Exit Stair 53

Table 4.23B Calculation of Present Value Net Savings of Occupant Evacuation Elevators vis-à-vis an Additional Exit Stair 54

Table 4.24A Calculation of Savings-to-Investment Ratio of Occupant Evacuation Elevators vis-à-vis an Additional Exit Stair 55
Table 4.24B Calculation of Savings-to-Investment Ratio of Occupant Evacuation Elevators vis-à-vis an Additional Exit Stair 56
Table 4.25A Calculation of Adjusted Internal Rate of Return of Occupant Evacuation Elevators vis-à-vis an Additional Exit Stair 57
Table 4.25B Calculation of Adjusted Internal Rate of Return of Occupant Evacuation Elevators vis-à-vis an Additional Exit Stair 58
Table 4.26 Assumptions for the Monte Carlo Simulations 60
Table 4.27A Summary Statistics from the Monte Carlo Simulations of Life-Cycle Cost for an Additional Exit Stair and Occupant Evacuation Elevators (statistics in thousands of dollars) 63
Table 4.27B Summary Statistics from the Monte Carlo Simulations of Life-Cycle Cost for an Additional Exit Stair and Occupant Evacuation Elevators (statistics in thousands of dollars) 63
Table 4.28A Summary Statistics of the Economic Performance from the Monte Carlo Simulations of Occupant Evacuation Elevators compared to an Additional Exit Stair (PVNS in thousands of dollars) 67
Table 4.28B Summary Statistics of the Economic Performance from the Monte Carlo Simulations of Occupant Evacuation Elevators compared to an Additional Exit Stair (PVNS in thousands of dollars) 68
Table 4.29A Discount Rate and Rental Rate Required for the Investment into Occupant Evacuation Elevators to Break-Even Over the Life-Cycle 69
Table 4.29B Discount Rate and Rental Rate Required for the Investment into Occupant Evacuation Elevators to Break-Even Over the Life-Cycle 70

List of Exhibits

Exhibit ES.1 Summary of Building 2 Cost-Effectiveness Analysis xx
Exhibit ES.2 Summary of Building 3 Cost-Effectiveness Analysis xxii
Exhibit ES.3 Summary of Building 4 Cost-Effectiveness Analysis xxiv
Exhibit ES.4 Summary of Building 5 Cost-Effectiveness Analysis xxvi
Exhibit 3.1 Format for Summarizing the Results of an Economic Analysis 22
Exhibit 4.1 Summary of the Building 2 Cost-Effectiveness Analysis 72
Exhibit 4.2 Summary of the Building 3 Cost-Effectiveness Analysis 74
Exhibit 4.3 Summary of the Building 4 Cost-Effectiveness Analysis 76
Exhibit 4.4 Summary of the Building 5 Cost-Effectiveness Analysis 78

Acronyms and Abbreviations

AIRR	Adjusted Internal Rate of Return
BC	Base Case
BFRL	Building and Fire Research Laboratory
BOMA	Building Owners and Managers Association
CBECS	Commercial Building Energy Consumption Survey
DOE	Department of Energy
GDP	Gross Domestic Product
GSA	General Service Administration
IBC	International Building Code
LCC	Life-Cycle Cost
NFPA	National Fire Protection Association
NIST	National Institute of Standards and Technology
OEES	Occupant Evacuation Elevator System
OMB	Office of Management Budget
PLEPM	Photoluminons Exit Path Markings
PVNS	Present Value Net Savings
RECS	Residential Energy Consumption Survey
SIR	Savings-to-Investment Ratio

Executive Summary

Fire protection measures are needed to maintain the safety and integrity of the Nation's building stock and to limit loss of life and property when building fires do occur. Statistics published by the National Fire Protection Association (NFPA) demonstrate that fire protection is a major investment cost in building construction. Therefore, ways to reduce the costs of fire protection while ensuring safety are of interest to building owners, fire protection engineers, and other construction industry stakeholders. Fire protection measures include, but are not limited to, building safety features concerned with extinguishment (e.g., sprinklers), containment (e.g., compartmentation), passive resistance (e.g., fire resistive materials), detection and alarm (e.g., smoke detectors), and egress (e.g., exit stairs). Although all fire protection measures have important economic implications, both in terms of first costs and future costs associated with operations and maintenance, the focus of this report is on egress-related measures in new building construction.

Egress-related measures are a major component of any fire protection strategy in buildings. Historically, building egress systems have evolved in response to specific large loss incidents. Aggressive building designs, changing occupant demographics, and consumer demand for more efficient systems have forced egress designs beyond the traditional exit stair-based approaches. Unfortunately, these approaches often lack a technical foundation for performance and economic trade-offs. Such performance and economic trade-off issues include the design for full-building and phased-evacuation of occupants, provisions for the evacuation of individuals with disabilities, and counterflow issues between first responders accessing a building and occupants evacuating a building.

The U.S. General Services Administration (GSA) commissioned this study to conduct an economic analysis of the use of elevators and exit stairs for occupant evacuation and fire service access for buildings greater than 120 ft (37 m) in height. GSA's general approach in the construction of new facilities and renovation projects in existing buildings is to incorporate cost-effective fire protection and life safety systems that result in overall building safety that meets or exceeds the levels required by local building codes. Unfortunately, there currently is a lack of data on the cost of existing, as well as newly proposed code requirements for occupant evacuation and fire service access in buildings. By sponsoring scientifically-based fire safety research, GSA will ensure building code requirements are cost-effective while ensuring safety. In addition, research will provide opportunities for GSA to evaluate the benefits of new technologies in their buildings and provide an opportunity to evaluate the cost of proposed code changes and ensure that they do not significantly increase construction and maintenance costs over the life of the building without demonstrably improving building safety.

Egress-related measures entail significant investment costs. In addition to initial capital investments, referred to as first costs, egress-related measures may result in significant future costs associated with major replacements as well as operations and maintenance costs. In some cases, egress-related measures may impinge on rentable floorspace, thus resulting in lost rental income. Therefore, any economic analyses must go beyond an evaluation of first cost considerations, because an alternative with higher first cost but lower future costs may be the most cost-effective choice. As a result, this report employs a life-cycle cost approach based on ASTM Standard Practice E 917 to analyze the costs of selected egress-related requirements. Where appropriate, additional cost related measures of economic performance are calculated.

Recent changes in the 2009 International Building Code (IBC) have set the stage for analyzing the costs of several key egress-related requirements. Four changes to the IBC that are examined in this report are:

- An additional exit stairway for buildings more than 420 ft (128 m) high;

- An increase of 50 percent in the width of exit stairways in new sprinklered buildings with floor area exceeding 15 000 ft^2 (1394 m^2);

- Permitting the use of elevators for occupant evacuation in fires and other emergencies for all buildings, and as an alternative to the required addition of an exit stairway for buildings more than 420 ft (128 m) high; and

- A minimum of one fire service access elevator for buildings more than 120 ft (37 m) high.

The purpose of this report is to produce economic analyses of cost data suitable for evaluating improved egress system designs that promote efficient and timely egress of occupants, including those with disabilities, and that facilitate more efficient fire department operations. This report tabulates cost data for selected egress-related requirements in five prototypical buildings. The five prototypical buildings range in height from a 5-floor, mid-rise building to a 75-floor, high-rise building. Cost data are tabulated in a format that facilitates life-cycle cost analyses of selected egress-related requirements. Incremental costs are also tabulated to help assess the implications of changing one or more design parameters.

This report contains five chapters and two appendices. Chapter 1 provides background information and introduces the subject matter. Chapter 2 provides a snapshot of the U.S. construction industry. Historical data on the value of construction put in place are used to highlight the NFPA procedure for estimating the annual cost of fire protection in buildings. Information from the U.S. Department of Energy's Commercial Buildings Energy Consumption Survey is then used to summarize the characteristics of commercial buildings, with special emphasis on buildings over 10 floors. For the most part, these buildings are more than 120 ft (37 m) high.

The methodology employed in this report to measure the economic performance of selected egress-related requirements is described in Chapter 3. The methodology is based on two types of analysis, four measures of economic performance, and a format for summarizing the results of an economic analysis of egress-related costs. The two types of analysis are baseline analysis and sensitivity analysis. The four measures of economic performance are life-cycle cost (LCC), present value of net savings (PVNS), savings-to-investment ratio (SIR), and adjusted internal rate of return (AIRR); they are based on Standard Practices issued by ASTM International. The format for summarizing the results of an economic analysis of egress-related costs is based on an ASTM Standard Guide.

Chapter 4 presents information on the five prototypical buildings used to derive cost data for selected egress-related requirements. Chapter 4 covers the central theme of this report and, as such, will be summarized in greater detail than the other sections of this report.

Chapter 5 concludes the report with a summary and a recommendation for further research. Specifically, the current procedure for estimating the national cost of fire protection in buildings needs to be revised to incorporate changes in codes, standards, practices, and technologies that have been adopted over the last 20 to 30 years. Once completed, the revised procedure will provide a comprehensive, science-based approach for estimating the national cost of fire protection in buildings.

Appendix A documents the specifications, assumptions, and cost estimating relationships used to develop costs of the occupant evacuation elevator systems for the four prototypical buildings over 120 ft (37 m) high presented in Chapter 4. Appendix B documents the specifications, assumptions, and cost estimating relationships used to develop the costs of the fire service access elevator systems for the four prototypical buildings over 120 ft (37 m) high presented in Chapter 4.

An objective of Chapter 4 is analyzing the cost-effectiveness of the required installation of an additional exit stair in buildings over 120 ft (37 m) high versus the alternative of installing an occupant evacuation elevator system. Characteristics of the five prototypical buildings are summarized in Table ES.1. Egress-related cost data were compiled from a number of sources, including industry experts, design professionals, and cost estimating guidebooks and software. Although the life-cycle costs of either an additional exit stair or an occupant evacuation elevator system are likely to amount to several million dollars (see Table ES.1), they represent only a small fraction (i.e., about 0.6 % to 1.2 %) of the construction cost of those buildings.

Within Chapter 4, cost data on exit stairs are first presented along with information on the incremental costs of increasing the width of exit stairs and the implications of installing an additional exit stair for buildings over 120 ft (37 m) high. Three nominal stair widths are considered: (1) 44 in (112 cm), the current minimum; (2) 56 in (142 cm), a width based on previous egress-related studies; and (3) 66 in (168 cm), the new minimum requirement for new sprinklered buildings with floor area exceeding 15 000 ft^2 (1394 m^2). Cost data on occupant evacuation elevator systems, and fire service access elevator systems, with special emphasis on buildings over 120 ft (37 m) high are then presented.

Table ES.1 Summary Information on the Prototypical Buildings Used in Developing Egress-Related Cost Data

Building	Number of Floors	Building Height Feet (ft)	Per Floor Area ft^2	Total Floorspace ft^2	Cost $ million	Cost $/ft^2
1	5	60	20,000	100,000	10.0	100.0
2	13	156	25,000	325,000	42.3	130.2
3	28	336	30,000	840,000	147.0	175.0
4	42	504	40,000	1,680,000	504.0	300.0
5	75	900	45,000	3,375,000	1,215.0	360.0

The life-cycle cost analysis presented in this report uses 2007 as the base year, a 25-year study period, and a 2.7 % real discount rate. The 2.7 % real discount rate conforms to Office of Management and Budget guidance for cost-effectiveness analyses of government programs with either a 20-year or 30-year study period. The length of the study period is based on the expected service life of elevators reported by Whitestone Research.

The baseline value for the life-cycle costs of installing an additional exit stair in Building 2 ranges from $1.5 million for the 44 in (112 cm) stair width to $2.4 million for the 66 in (168 cm) stair width. The baseline value for the life-cycle costs of installing an additional exit stair in Building 3 ranges from $3.2 million for the 44 in (112 cm) stair width to $5.2 million for the 66 in (168 cm) stair width. The baseline value for the life-cycle costs of installing an additional exit stair in Building 4 ranges from $4.8 million for the 44 in (112 cm) stair width to $7.8 million for the 66 in (168 cm) stair width. The baseline value for the life-cycle costs of installing an additional exit stair in Building 5 ranges from $8.6 million for the 44 in (112 cm) stair width to $13.9 million for the 66 in (168 cm) stair width.

The baseline value for the life-cycle costs of converting a standard passenger elevator system to an occupant evacuation elevator system in Building 2 is $0.5 million. The baseline value for the life-cycle costs of converting a standard passenger elevator system to an occupant evacuation elevator system in Building 3 is $1.1 million. The baseline value for the life-cycle costs of converting a standard passenger elevator system to an occupant evacuation elevator system in Building 4 is $2.4 million. The baseline value for the life-cycle costs of converting a standard passenger elevator system to an occupant evacuation elevator system in Building 5 is $7.6 million.

It is important to note that although the life-cycle cost of converting a standard passenger elevator system to an occupant evacuation elevator system is less than the life-cycle cost of installing an additional exit stair in Buildings 2 through 5, the first cost of converting a standard passenger elevator system to an occupant evacuation elevator system is higher. Thus, the installation of an exit stair is designated as the Base Case—because it has a lower first cost—when performing the cost-effectiveness analysis.

Baseline values for the life-cycle costs of converting a standard service elevator system to a fire service access elevator system are calculated for Buildings 2 through 5, the four prototypical buildings over 120 ft (37 m) high. These values are: $195 thousand for Building 2; $395 thousand for Building 3; $660 thousand for Building 4; and $1.7 million for Building 5.

Because the values of many variables that enter into the baseline analysis are not known with certainty, it is advisable to select a small set of variables whose impact is likely to be substantial and subject them to a sensitivity analysis. Five inputs into the baseline analysis are identified as containing uncertainty in their estimates. The five inputs are: (1) the discount rate; (2) the rental rate for commercial office space; (3) the cost of the fire door and frame assembly protecting lobbies above the ground floor; (4) the Sky Lobby lost rental space for Building 5; and (5) the cost associated with signage, lobby status indicator, and a two-way communication system (see Table ES.2). The discount rate affects the life-cycle costs of both alternatives—installing an additional exit stair or installing occupant evacuation elevators—in Buildings 2 through 5. The rental rate affects the life-cycle costs of installing an additional exit stair in Buildings 2 through 5 and installing occupant evacuation elevators in Building 5. The costs associated with the fire door and frame assembly and costs associated with signage, lobby status indicator, and a two-way communication system affect the life-cycle cost of installing occupant evacuation elevators in Buildings 2 through 5. The Sky Lobby lost rental space affects the life-cycle cost of installing occupant evacuation elevators in Building 5.

Table ES.2 Assumptions for the Monte Carlo Simulations

Variable	Probability Distribution	Setting & Value		
		Min	Most-Likely	Max
Discount Rate	Triangular	1.0	2.7	10.0
Rental Rate	Triangular	$ 18.18	$ 36.92	$ 43.18
Fire Door and Frame System	Triangular	$ 4,000	$ 5,250	$ 6,000
Sky Lobby Lost Rental Space	Triangular	-	4,000	8,403
Signage, Lobby Status Indicator, and Two-Way Communication System	Triangular	$ 4,000	$ 4,500	$ 5,000

Monte Carlo simulation is used to assess the sensitivity of the life-cycle costs of an additional exit stair and the occupant evacuation elevator system to changes in the five uncertainty parameters. Each input is designed to follow a triangular distribution. Throughout this sensitivity analysis, 10 000 simulations were run for each combination of input variables under analysis.

The results of the baseline analysis and the sensitivity analysis are summarized in Exhibits ES.1 (Building 2), ES.2 (Building 3), ES.3 (Building 4), and ES.4 (Building 5). The summary format is based on ASTM Standard Guide E 2204.

Exhibit ES.1 Summary of Building 2 Cost-Effectiveness Analysis

1.a Significance of the Project:
Recent changes to the International Building Code (IBC) affect egress-related measures in buildings over 420 ft (128 m) high. GSA was interested in evaluating the changes for buildings over 120 ft (37 m) in height. Two such changes—an additional exit stair and permitting the use of occupant evacuation elevators as an alternative to the required addition of an exit stair—were the focus of an economic analysis. Information on the costs and specifications of alternative configurations for exit stairs and occupant evacuation elevators were compiled to support the economic analysis and to serve as a resource for building owners, fire protection engineers, and key construction industry stakeholders concerned about egress and life-safety issues in high rise buildings.

The economic analysis was commissioned by GSA in support of its objective to incorporate cost-effective fire protection and life safety systems that result in overall building safety that meets or exceeds the levels required by local building codes. The economic analysis was conducted in two phases. First a baseline analysis was performed holding all input variables at their most likely values. Second a sensitivity analysis employing Monte Carlo simulation was performed through which probabilistic levels of significance were calculated for the key measures of economic performance. The results of the economic analysis demonstrate that occupant evacuation elevators are a cost-effective alternative to the installation of an additional exit stair. Furthermore, these results are fairly robust, as demonstrated in the sensitivity analysis where key input variables were varied about their baseline values.

1.b Key Points:

- Recent changes to the IBC affect egress-related measures in buildings over 420 ft (128 m) high.
- GSA commissioned NIST to perform an economic analysis in support of its objective to incorporate cost-effective fire protection and life safety systems that result in overall building safety that meets or exceeds the levels required by local building codes.
- GSA was interested in evaluating the changes for buildings over 120 ft (37 m) in height.
- NIST compiled cost data on alternative configurations for exit stairs and occupant evacuation elevators to support the economic analysis and to serve as an information resource for building owners, fire protection engineers, and other key construction industry stakeholders concerned about egress and life-safety issues in high rise buildings.
- The results of the economic analysis demonstrate that occupant evacuation elevators are a cost-effective alternative to the installation of an additional exit stair.

2. Analysis Strategy: How Key Measures are Estimated

The following economic measures are calculated as present-value (PV) amounts:
(1) **Life-Cycle Costs** (LCC) for the Base Case (Additional Exit Stair of nominal width 44 in (112 cm), 56 in (142 cm), and 66 in (168 cm)) and for the Proposed Alternative (Occupant Evacuation Elevators), including all costs of installing and operating the two systems over the length of the study period. The selection criterion is lowest LCC.
(2) **Present Value Net Savings** (PVNS) that will result from selecting the lowest-LCC alternative. PVNS > 0 indicates an economically worthwhile project.

Additional measures:
(1) **Savings-to-Investment Ratio** (SIR), the ratio of savings from the lowest-LCC to the extra investment required to implement it. A ratio of SIR >1 indicates an economically worthwhile project.
(2) **Adjusted Internal Rate of Return** (AIRR), the annual return on investment over the study period. An AIRR > discount or hurdle rate indicates an economically worthwhile project.

Data and Assumptions:
- The Base Date is 2007.
- The alternative with the lower first cost (Additional Exit Stair) is designated the Base Case.
- The study period is 25 years and ends in 2031.
- The baseline value of the discount or hurdle rate is 2.7 % real.

| 3.a Calculation of Savings, Costs, and Additional Measures ||||||
| Results of Baseline Analysis (Savings and Costs in Thousands of Dollars) ||||||
Economic Measure	Base Case (44)	OEES	Base Case (66)	OEES	
Life-Cycle Cost (LCC)	$1,501	$551	$2,402	$551	
Investment Cost	$114	$453	$48	$453	
Delta Investment Cost	N/A	$339	N/A	$405	
Non-Investment Cost	$1,387	$97	$2,354	$97	
Savings	N/A	$1,290	N/A	$2,257	
Present Value Net Savings (PVNS)	N/A	$950	N/A	$1,827	
Savings-to-Investment Ratio (SIR)	N/A	3.80	N/A	5.57	
AIRR	N/A	8.33 %	N/A	10.00 %	

| Results of Monte Carlo Simulation (Savings and Costs in Thousands of Dollars) ||||||
| Economic Measure | Statistical Measure |||||
	Minimum	Median	Maximum	Mean	Standard Deviation
$LCC_{BC(44)}$	495	1,134	1,949	1,150	260
$LCC_{BC(56)}$	601	1,469	2,578	1,490	353
$LCC_{BC(66)}$	695	1,779	3,163	5,832	440
LCC_{OEES}	447	527	601	527	26
$PVNS_{BC(44):OEES}$	-35	608	1,386	623	779
$PVNS_{BC(56):OEES}$	71	943	2,010	963	1,079
$PVNS_{BC(66):OEES}$	166	1,253	2,595	1,279	1,361

3.b Key Results

*LCC (Thousands of Dollars)

Base Case (44)	$1,501
Base Case (56)	$1,968
Base Case (66)	$2,402
OEES	$551

*PVNS (Thousands of Dollars)

BC(44):OEES	$951
BC(56):OEES	$1,418
BC(66):OEES	$1,852

*SIR

BC(44):OEES	3.80
BC(56):OEES	4.82
BC(66):OEES	5.57

*AIRR

BC(44):OEES	8.33 %
BC(56):OEES	9.37 %
BC(66):OEES	10.00 %

3.c Traceability

Life-cycle costs and supplementary measures were calculated according to ASTM standards E 917, E 964, E 1057, and E 1074. Treatment of uncertainty and measures of project risk were calculated according to ASTM standards E 1369 and E 1946.

Section 3008 of the 2009 edition of the International Building Code specifies the criteria that passenger elevators must meet to be used for evacuation purposes.

Exhibit ES.2 Summary of Building 3 Cost-Effectiveness Analysis

1.a Significance of the Project: Recent changes to the International Building Code (IBC) affect egress-related measures in buildings over 420 ft (128 m) high. GSA was interested in evaluating the changes for buildings over 120 ft (37 m) in height. Two such changes—an additional exit stair and permitting the use of occupant evacuation elevators as an alternative to the required addition of an exit stair—were the focus of an economic analysis. Information on the costs and specifications of alternative configurations for exit stairs and occupant evacuation elevators were compiled to support the economic analysis and to serve as a resource for building owners, fire protection engineers, and key construction industry stakeholders concerned about egress and life-safety issues in high rise buildings. The economic analysis was commissioned by GSA in support of its objective to incorporate cost-effective fire protection and life safety systems that result in overall building safety that meets or exceeds the levels required by local building codes. The economic analysis was conducted in two phases. First a baseline analysis was performed holding all input variables at their most likely values. Second a sensitivity analysis employing Monte Carlo simulation was performed through which probabilistic levels of significance were calculated for the key measures of economic performance. The results of the economic analysis demonstrate that occupant evacuation elevators are a cost-effective alternative to the installation of an additional exit stair. Furthermore, these results are fairly robust, as demonstrated in the sensitivity analysis where key input variables were varied about their baseline values.	**1.b Key Points:** • Recent changes to the IBC affect egress-related measures in buildings over 420 ft (128 m) high. • GSA commissioned NIST to perform an economic analysis in support of its objective to incorporate cost-effective fire protection and life safety systems that result in overall building safety that meets or exceeds the levels required by local building codes. • GSA was interested in evaluating the changes for buildings over 120 ft (37 m) in height. • NIST compiled cost data on alternative configurations for exit stairs and occupant evacuation elevators to support the economic analysis and to serve as an information resource for building owners, fire protection engineers, and other key construction industry stakeholders concerned about egress and life-safety issues in high rise buildings. • The results of the economic analysis demonstrate that occupant evacuation elevators are a cost-effective alternative to the installation of an additional exit stair.

2. Analysis Strategy: How Key Measures are Estimated

The following economic measures are calculated as present-value (PV) amounts:

(3) **Life-Cycle Costs** (LCC) for the Base Case (Additional Exit Stair of nominal width 44 in (112 cm), 56 in (142 cm), and 66 in (168 cm)) and for the Proposed Alternative (Occupant Evacuation Elevators), including all costs of installing and operating the two systems over the length of the study period. The selection criterion is lowest LCC.

(4) **Present Value Net Savings** (PVNS) that will result from selecting the lowest-LCC alternative. PVNS > 0 indicates an economically worthwhile project.

Additional measures:

(3) **Savings-to-Investment Ratio** (SIR), the ratio of savings from the lowest-LCC to the extra investment required to implement it. A ratio of SIR >1 indicates an economically worthwhile project.

(4) **Adjusted Internal Rate of Return** (AIRR), the annual return on investment over the study period. An AIRR > discount or hurdle rate indicates an economically worthwhile project.

Data and Assumptions:
- The Base Date is 2007.
- The alternative with the lower first cost (Additional Exit Stair) is designated the Base Case.
- The study period is 25 years and ends in 2031.
- The baseline value of the discount or hurdle rate is 2.7 % real.

3.a Calculation of Savings, Costs, and Additional Measures

Results of Baseline Analysis (Savings and Costs in Thousands of Dollars)

Economic Measure	Base Case (44)	OEES	Base Case (66)	OEES
Life-Cycle Cost (LCC)	$3,229	$1,403	$5,173	$1,403
Investment Cost	241	$1,143	$103	$1,143
Delta Investment Cost	N/A	$902	N/A	$1040
Non-Investment Cost	$2,988	$259	$5,070	$259
Savings	N/A	$2729	N/A	$4811
Present Value Net Savings (PVNS)	N/A	$1,827	N/A	$3,770
Savings-to-Investment Ratio (SIR)	N/A	3.02	N/A	4.62
AIRR	N/A	7.35 %	N/A	9.19 %

Results of Monte Carlo Simulation (Savings and Costs in Thousands of Dollars)

Economic Measure	Statistical Measure				
	Minimum	Median	Maximum	Mean	Standard Deviation
$LCC_{BC(44)}$	1,063	2,438	4,195	2,472	559
$LCC_{BC(56)}$	1,293	3,162	5,550	3,208	760
$LCC_{BC(66)}$	1,497	3,830	6,812	3,888	949
LCC_{OEES}	1,152	1,344	1,522	1,343	62
$PVNS_{BC(44):OEES}$	-274	1,100	2,762	1,129	1,234
$PVNS_{BC(56):OEES}$	-44	1,822	4,102	1,865	1,649
$PVNS_{BC(66):OEES}$	160	2,491	5,364	2,545	2,091

3.b Key Results

*LCC (Thousands of Dollars)

Base Case (44)	$3,229
Base Case (56)	$4,237
Base Case (66)	$5,173
OEES	$1,403

*PVNS (Thousands of Dollars)

BC(44):OEES	$1,827
BC(56):OEES	$2,835
BC(66):OEES	$3,770

*SIR

BC(44):OEES	3.02
BC(56):OEES	3.93
BC(66):OEES	4.62

*AIRR

BC(44):OEES	7.35 %
BC(56):OEES	8.48 %
BC(66):OEES	9.19 %

3.c Traceability

Life-cycle costs and supplementary measures were calculated according to ASTM standards E 917, E 964, E 1057, and E 1074. Treatment of uncertainty and measures of project risk were calculated according to ASTM standards E 1369 and E 1946.

Section 3008 of the 2009 edition of the International Building Code specifies the criteria that passenger elevators must meet to be used for evacuation purposes.

Exhibit ES.3 Summary of Building 4 Cost-Effectiveness Analysis

1.a Significance of the Project:	1.b Key Points:
Recent changes to the International Building Code (IBC) affect egress-related measures in buildings over 420 ft (128 m) high. Two such changes—an additional exit stair and permitting the use of occupant evacuation elevators as an alternative to the required addition of an exit stair—were the focus of an economic analysis. Information on the costs and specifications of alternative configurations for exit stairs and occupant evacuation elevators were compiled to support the economic analysis and to serve as a resource for building owners, fire protection engineers, and key construction industry stakeholders concerned about egress and life-safety issues in high rise buildings. The economic analysis was commissioned by GSA in support of its objective to incorporate cost-effective fire protection and life safety systems that result in overall building safety that meets or exceeds the levels required by local building codes. The economic analysis was conducted in two phases. First a baseline analysis was performed holding all input variables at their most likely values. Second a sensitivity analysis employing Monte Carlo simulation was performed through which probabilistic levels of significance were calculated for the key measures of economic performance. The results of the economic analysis demonstrate that occupant evacuation elevators are a cost-effective alternative to the installation of an additional exit stair. Furthermore, these results are fairly robust, as demonstrated in the sensitivity analysis where key input variables were varied about their baseline values.	• Recent changes to the IBC affect egress-related measures in buildings over 420 ft (128 m) high. • GSA commissioned NIST to perform an economic analysis in support of its objective to incorporate cost-effective fire protection and life safety systems that result in overall building safety that meets or exceeds the levels required by local building codes. • NIST compiled cost data on alternative configurations for exit stairs and occupant evacuation elevators to support the economic analysis and to serve as an information resource for building owners, fire protection engineers, and other key construction industry stakeholders concerned about egress and life-safety issues in high rise buildings. • The results of the economic analysis demonstrate that occupant evacuation elevators are a cost-effective alternative to the installation of an additional exit stair.

2. Analysis Strategy: How Key Measures are Estimated

The following economic measures are calculated as present-value (PV) amounts:

(5) **Life-Cycle Costs** (LCC) for the Base Case (Additional Exit Stair of nominal width 44 in (112 cm), 56 in (142 cm), and 66 in (168 cm)) and for the Proposed Alternative (Occupant Evacuation Elevators), including all costs of installing and operating the two systems over the length of the study period. The selection criterion is lowest LCC.

(6) **Present Value Net Savings** (PVNS) that will result from selecting the lowest-LCC alternative. PVNS > 0 indicates an economically worthwhile project.

Additional measures:

(5) **Savings-to-Investment Ratio** (SIR), the ratio of savings from the lowest-LCC to the extra investment required to implement it. A ratio of SIR >1 indicates an economically worthwhile project.

(6) **Adjusted Internal Rate of Return** (AIRR), the annual return on investment over the study period. An AIRR > discount or hurdle rate indicates an economically worthwhile project.

Data and Assumptions:
- The Base Date is 2007.
- The alternative with the lower first cost (Additional Exit Stair) is designated the Base Case.
- The study period is 25 years and ends in 2031.
- The baseline value of the discount or hurdle rate is 2.7 % real.

3.a Calculation of Savings, Costs, and Additional Measures					
Results of Baseline Analysis (Savings and Costs in Thousands of Dollars)					
Economic Measure		Base Case (44)	OEES	Base Case (66)	OEES
Life-Cycle Cost (LCC)		$4,842	$2,440	$7,759	$2,440
Investment Cost		$360	$1,921	$154	$1,921
Delta Investment Cost		N/A	$1,561	N/A	$1,767
Non-Investment Cost		$4,482	$519	$7,605	$519
Savings		N/A	$3,963	N/A	$7,086
Present Value Net Savings (PVNS)		N/A	$2,402	N/A	$5,319
Savings-to-Investment Ratio (SIR)		N/A	2.54	N/A	4.01
AIRR		N/A	6.60 %	N/A	8.57 %
Results of Monte Carlo Simulation (Savings and Costs in Thousands of Dollars)					
Economic Measure	Statistical Measure				
	Minimum	Median	Maximum	Mean	Standard Deviation
$LCC_{BC(44)}$	1,593	3,655	6,291	3,707	839
$LCC_{BC(56)}$	1,938	4,742	8,324	4,811	1,140
$LCC_{BC(66)}$	2,245	5,745	10,218	5,832	1,423
LCC_{OEES}	2,004	2,330	2,644	2,330	109
$PVNS_{BC(44):OEES}$	-687	1,337	3,792	1,377	780
$PVNS_{BC(56):OEES}$	-341	2,421	5,798	2,482	1,080
$PVNS_{BC(66):OEES}$	-34	3,425	7,691	3,502	1,362

3.b Key Results

*LCC (Thousands of Dollars)

Base Case (44)	$4,842
Base Case (56)	$6,355
Base Case (66)	$7,759
OEES	$2,440

*PVNS (Thousands of Dollars)

BC(44):OEES	$2,402
BC(56):OEES	$3,915
BC(66):OEES	$5,319

*SIR

BC(44):OEES	2.54
BC(56):OEES	3.36
BC(66):OEES	4.01

*AIRR

BC(44):OEES	6.60 %
BC(56):OEES	7.80 %
BC(66):OEES	8.57 %

3.c Traceability

Life-cycle costs and supplementary measures were calculated according to ASTM standards E 917, E 964, E 1057, and E 1074. Treatment of uncertainty and measures of project risk were calculated according to ASTM standards E 1369 and E 1946.

Section 3008 of the 2009 edition of the International Building Code specifies the criteria that passenger elevators must meet to be used for evacuation purposes.

Exhibit ES.4 Summary of Building 5 Cost-Effectiveness Analysis

1.a Significance of the Project: Recent changes to the International Building Code (IBC) affect egress-related measures in buildings over 420 ft (128 m) high. Two such changes—an additional exit stair and permitting the use of occupant evacuation elevators as an alternative to the required addition of an exit stair—were the focus of an economic analysis. Information on the costs and specifications of alternative configurations for exit stairs and occupant evacuation elevators were compiled to support the economic analysis and to serve as a resource for building owners, fire protection engineers, and key construction industry stakeholders concerned about egress and life-safety issues in high rise buildings. The economic analysis was commissioned by GSA in support of its objective to incorporate cost-effective fire protection and life safety systems that result in overall building safety that meets or exceeds the levels required by local building codes. The economic analysis was conducted in two phases. First a baseline analysis was performed holding all input variables at their most likely values. Second a sensitivity analysis employing Monte Carlo simulation was performed through which probabilistic levels of significance were calculated for the key measures of economic performance. The results of the economic analysis demonstrate that occupant evacuation elevators are a cost-effective alternative to the installation of an additional exit stair. Furthermore, these results are fairly robust, as demonstrated in the sensitivity analysis where key input variables were varied about their baseline values.	**1.b Key Points:** - Recent changes to the IBC affect egress-related measures in buildings over 420 ft (128 m) high. - GSA commissioned NIST to perform an economic analysis in support of its objective to incorporate cost-effective fire protection and life safety systems that result in overall building safety that meets or exceeds the levels required by local building codes. - NIST compiled cost data on alternative configurations for exit stairs and occupant evacuation elevators to support the economic analysis and to serve as an information resource for building owners, fire protection engineers, and other key construction industry stakeholders concerned about egress and life-safety issues in high rise buildings. - The results of the economic analysis demonstrate that occupant evacuation elevators are a cost-effective alternative to the installation of an additional exit stair.

2. Analysis Strategy: How Key Measures are Estimated

The following economic measures are calculated as present-value (PV) amounts:

(7) **Life-Cycle Costs** (LCC) for the Base Case (Additional Exit Stair of nominal width 44 in (112 cm), 56 in (142 cm), and 66 in (168 cm)) and for the Proposed Alternative (Occupant Evacuation Elevators), including all costs of installing and operating the two systems over the length of the study period. The selection criterion is lowest LCC.

(8) **Present Value Net Savings** (PVNS) that will result from selecting the lowest-LCC alternative. PVNS > 0 indicates an economically worthwhile project.

Additional measures:

(7) **Savings-to-Investment Ratio** (SIR), the ratio of savings from the lowest-LCC to the extra investment required to implement it. A ratio of SIR >1 indicates an economically worthwhile project.

(8) **Adjusted Internal Rate of Return** (AIRR), the annual return on investment over the study period. An AIRR > discount or hurdle rate indicates an economically worthwhile project.

Data and Assumptions:
- The Base Date is 2007.
- The alternative with the lower first cost (Additional Exit Stair) is designated the Base Case.
- The study period is 25 years and ends in 2031.
- The baseline value of the discount or hurdle rate is 2.7 % real.

| 3.a Calculation of Savings, Costs, and Additional Measures ||||||
| Results of Baseline Analysis (Savings and Costs in Thousands of Dollars) ||||||
Economic Measure	Base Case (44)	OEES	Base Case (66)	OEES	
Life-Cycle Cost (LCC)	$8,644	$7,604	$13,855	$7,604	
Investment Cost	$640	$3,761	$274	$3,761	
Delta Investment Cost	N/A	$3,121	N/A	$3,487	
Non-Investment Cost	$8,004	$3,843	$13,581	$3,843	
Savings	N/A	$4,161	N/A	$9,738	
Present Value Net Savings (PVNS)	N/A	$1,040	N/A	$6,251	
Savings-to-Investment Ratio (SIR)	N/A	1.33	N/A	2.79	
AIRR	N/A	3.89 %	N/A	7.01 %	
Results of Monte Carlo Simulation (Savings and Costs in Thousands of Dollars)					
Economic Measure	Statistical Measure				
	Minimum	Median	Maximum	Mean	Standard Deviation
$LCC_{BC(44)}$	2,841	6,525	11,232	6,616	1,498
$LCC_{BC(56)}$	3,459	8,466	14,864	8,590	2,036
$LCC_{BC(66)}$	4,009	10,258	18,245	10,414	2,542
LCC_{OEES}	4,271	6,641	11,408	6,768	1,114
$PVNS_{BC(44):OEES}$	-2,880	-301	5,876	-151	1,235
$PVNS_{BC(56):OEES}$	-2,148	1,639	9,410	1,822	1,651
$PVNS_{BC(66):OEES}$	-1,599	3,446	12,708	3,646	2,093

3.b Key Results

*LCC (Thousands of Dollars)

Base Case (44)	$8,644
Base Case (56)	$11,347
Base Case (66)	$13,855
OEES	$7,604

*PVNS (Thousands of Dollars)

BC(44):OEES	$1,040
BC(56):OEES	$3,743
BC(66):OEES	$6,251

*SIR

BC(44):OEES	1.33
BC(56):OEES	2.14
BC(66):OEES	2.79

*AIRR

BC(44):OEES	3.89 %
BC(56):OEES	5.87 %
BC(66):OEES	7.01 %

3.c Traceability

Life-cycle costs and supplementary measures were calculated according to ASTM standards E 917, E 964, E 1057, and E 1074. Treatment of uncertainty and measures of project risk were calculated according to ASTM standards E 1369 and E 1946.

Section 3008 of the 2009 edition of the International Building Code specifies the criteria that passenger elevators must meet to be used for evacuation purposes.

Because the format is fairly compact, it is necessary to abbreviate some of the terms reported in the exhibits. The term Base Case is used to represent the installation of an additional exit stair because the first cost for each of the three exit stair configurations was lower than the first cost for an occupant evacuation elevator system. The abbreviation BC refers to the Base Case (exit stair). The values in parentheses—(44), (56), and (66)—refer to the width of the additional exit stair. The abbreviation OEES refers to the occupant evacuation elevator system. The abbreviations BC(44), BC(56), BC(66), and OEES refer to the corresponding exit stair configurations and the occupant evacuation elevator system. For example, $LCC_{BC(44)}$ corresponds to the life-cycle cost of the 44 in (112 cm) wide exit stair. The abbreviations BC(44):OEES, BC(56):OEES, and BC(66):OEES are used to represent comparisons between a given exit stair width and the corresponding occupant evacuation elevator system. For example, $PVNS_{BC(66):OEES}$ corresponds to the present value net savings of the occupant evacuation elevator system vis-à-vis the 66 in (168 cm) wide exit stair.

The results of the baseline analysis, reported in the exhibits, demonstrate that occupant evacuation elevators are a cost-effective alternative to the installation of an additional exit stair for the four prototypical buildings over 120 ft (37 m) high. Furthermore, these results are reasonably representative, as demonstrated in the sensitivity analysis, with some notable exceptions.

While installation of occupant evacuation elevators is generally cost-effective vis-à-vis the installation of an additional exit stair on a life-cycle cost basis, the sensitivity analysis reveals combinations of the uncertainty parameters which produce comparisons that are not. This is best seen by referring to the entries for the present value of net savings (PVNS) entries in the exhibits. A negative PVNS value indicates that installation of occupant evacuation elevators are not cost effective vis-à-vis the installation of an additional exit stair.

For Building 2, PVNS ranges from a minimum of -$0.04 million (compared to the 44 in (112 cm) exit stair) to a maximum of $2.6 million (compared to the 66 in (168 cm) exit stair). Refer next to the negative PVNS entries in Exhibit ES.1. For the comparison with the 44 in (112 cm) exit stair, a negative PVNS occurred in 0.02 % of the simulations. For the comparison with the 56 in (142 cm) exit stair and 66 in (168 cm) exit stair, the occupant evacuation elevators were cost-effective. For all three exit stair widths, minimum PVNS corresponds with low rental rates and high discount rates.

For Building 3, PVNS ranges from a minimum of -$0.3 million (compared to the 44 in (112 cm) exit stair) to a maximum of $5.4 million (compared to the 66 in (168 cm) exit stair). Refer next to the negative PVNS entries in Exhibit ES.2. For the comparison with the 44 in (112 cm) exit stair, a negative PVNS occurred in 0.5 % of the simulations. For the comparison with the 56 in (142 cm) exit stair, a negative PVNS occurred in 0.02 % of the simulations. For the comparison with the 66 in (168 cm) exit stair, the occupant evacuation elevator was cost-effective. For all three exit stair widths, minimum PVNS corresponds with low rental rates and high discount rates.

For Building 4, PVNS ranges from a minimum of -$0.7 million (compared to the 44 in (112 cm) exit stair) to a maximum of $7.7 million (compared to the 66 in (168 cm) exit stair). Refer next to the negative PVNS entries in Exhibit ES.3. For the comparison with the 44 in (112 cm) exit stair, a negative PVNS occurred in 2.6 % of the simulations. For the comparison with the 56 in (142 cm) exit stair, a negative PVNS occurred in 0.1 % of the simulations. For the comparison with the 66 in (168 cm) exit stair, a negative PVNS occurred in 0.01 % of the simulations. For all three exit stair widths, minimum PVNS corresponds with low rental rates and high discount rates.

For Building 5, PVNS ranges from a minimum of -$2.9 million (compared to the 44 in (112 cm) exit stair) to a maximum of $12.7 million (compared to the 66 in (168 cm) exit stair). Refer next to the negative PVNS entries in Exhibit ES.4. For the comparison with the 44 in (112 cm) exit stair, a negative PVNS occurred in 59.2 % of the simulations. For the comparison with the 56 in (142 cm) exit stair, a negative PVNS occurred in 13.0 % of the simulations. For the comparison with the 66 in (168 cm) exit stair, a negative PVNS occurred in 2.1 % of the simulations. For all three stair widths, minimum PVNS corresponds with low rental rates, large Sky Lobby areas, and high discount rates. Large Sky Lobby areas affect both lobby enclosure costs, a first cost, and rentable floorspace on the Sky Lobby floor, an annual recurring cost.

In summary, the results of the economic analysis for the four prototypical buildings over 120 ft (37 m) high, demonstrate that: (1) an additional exit stair is a cost-effective alternative to the installation of occupant evacuation elevators on a first-cost basis; and (2) occupant evacuation elevators are a cost-effective alternative to the installation of an additional exit stair on a life-cycle cost basis when rental rates are high and discount rates are low.

1 Introduction

1.1 Background

Fire protection measures are needed to maintain the safety and integrity of the Nation's building stock and to limit loss of life and property when building fires do occur. Statistics published by the National Fire Protection Association (NFPA) demonstrate that fire protection is a major investment cost in building construction.[1] Therefore, ways to reduce the costs of fire protection while ensuring safety are of interest to building owners, fire protection engineers, and other construction industry stakeholders. Fire protection measures include, but are not limited to, building safety features concerned with extinguishment (e.g., sprinklers), containment (e.g., compartmentation), passive resistance (e.g., fire resistive materials), detection and alarm (e.g., smoke detectors), and egress (e.g., exit stairs). Although all fire protection measures have important economic implications, both in terms of first costs and future costs associated with operations and maintenance, the focus of this report is on egress-related measures in new building construction.

Egress-related measures are a major component of any fire protection strategy in buildings. Historically, building egress systems have evolved in response to specific large loss incidents. Aggressive building designs, changing occupant demographics, and consumer demand for more efficient systems have forced egress designs beyond the traditional exit stair-based approaches. Unfortunately, these approaches often lack a technical foundation for performance and economic trade-offs. Such performance and economic trade-off issues include the design for full-building and phased-evacuation of occupants, provisions for the evacuation of individuals with disabilities, and counterflow issues between first responders accessing a building and occupants evacuating a building.

The U.S. General Services Administration (GSA) commissioned this study to conduct an economic analysis of the use of elevators and exit stairs for occupant evacuation and fire service access for buildings greater than 120 ft (37 m) in height. GSA's general approach in the construction of new facilities and renovation projects in existing buildings is to incorporate cost-effective fire protection and life safety systems that result in overall building safety that meets or exceeds the levels required by local building codes. Unfortunately, there currently is a lack of data on the cost of existing, as well as newly proposed code requirements for occupant evacuation and fire service access in buildings. By sponsoring scientifically-based fire safety research, GSA will ensure building code requirements are cost-effective while ensuring safety. In addition, research will provide opportunities for GSA to evaluate the benefits of new technologies in their buildings and provide an opportunity to evaluate the cost of proposed code changes and ensure that they do not significantly increase construction and maintenance costs over the life of the building without demonstrably improving building safety.

---

[1] Hall, J.R. 2008. "The Total Cost of Fire in the United States." Quincy, MA: National Fire Protection Association.

Egress-related measures entail significant investment costs. In addition to initial capital investments, referred to as first costs, egress-related measures may result in significant future costs associated with major replacements as well as operations and maintenance costs. In some cases, egress-related measures may impinge on rentable floorspace, thus resulting in lost rental income. Therefore, any economic analyses must go beyond an evaluation of first cost considerations, because an alternative with higher first cost but lower future costs may be the most cost-effective choice. As a result, this report employs a life-cycle cost approach based on ASTM Standard Practice E 917[2] to analyze the costs of selected egress-related requirements. Where appropriate, additional cost related measures of economic performance are calculated.

Recent changes in the International Building Code (IBC)[3] have set the stage for analyzing the costs of several key egress-related requirements.[4] Four changes to the IBC that are examined in this report are:

- An additional exit stairway for buildings more than 420 ft (128 m) high;

- An increase of 50 percent in the width of exit stairways in new sprinklered buildings with floor area exceeding 15 000 ft^2 (1394 m^2);[5]

- Permitting the use of elevators for occupant evacuation in fires and other emergencies for all buildings, and as an alternative to the required addition of an exit stairway for buildings more than 420 ft (128 m) high;[6] and

- A minimum of one fire service access elevator for buildings more than 120 ft (37 m) high.[7]

---

[2] For a detailed description of the ASTM life-cycle cost standard, see ASTM International. "Standard Practice for Measuring Life-Cycle Costs of Buildings and Building Systems," E 917, *Annual Book of ASTM Standards: 2008*, Vol. 04.11. West Conshohocken, PA: ASTM International.
[3] International Code Council, Inc. 2009. *International Building Code*. Washington, DC: International Code Council, Inc.
[4] National Institute of Standards and Technology. "Safer Buildings Are Goal of New Code Changes Based on Recommendations from NIST World Trade Center Investigation" TechBeat: October 1, 2008. http://www.nist.gov/public_affairs/releases/wtc_100108.html (accessed December 2008).
[5] Due to changes to Section 1005 of the IBC. See *NIST's Recommendations Following the Federal Building and Fire Investigation of the World Trade Center Disaster*, http://wtc.nist.gov/recommendations/recommendations.htm#Recommendation_17.
It states,
> Requires that the width of exit stairs in all (new) buildings be calculated on the basis that has traditionally been applied to unsprinklered buildings, which is 50 percent greater than what was permitted for sprinklered buildings. This will result in wider exit stairs where the occupant load exceeds the minimum capacity provided by two 1100 mm (44 in) stairs. For an office building, wider stairs will be required in floors with a gross area exceeding about 15,000 sq ft per stair (30,000 sq ft with the minimum of two 1100 mm stairs).

[6] Passenger elevators must meet specific criteria to be used for occupant evacuation purposes; these criteria are provided in Section 3008 of the *2009 International Building Code*, Op Cit.
[7] Elevators must meet specific criteria to be used for fire service access; these criteria are provided in Section 3007 of the *2009 International Building Code*, Ibid.

1.2 Purpose

The purpose of this report is to produce economic analyses of cost data suitable for evaluating improved egress system designs that promote efficient and timely egress of occupants, including those with disabilities, and that facilitate more efficient fire department operations. This report tabulates cost data for selected egress-related requirements in five prototypical buildings specified by GSA. The five prototypical buildings range in height from a 5-floor, mid-rise building to a 75-floor, high-rise building. Cost data are tabulated in a format that facilitates life-cycle cost analyses of selected egress-related requirements. Incremental costs are also tabulated to help assess the implications of changing one or more design parameters.

1.3 Scope and Approach

This report contains four chapters and two appendices in addition to the Introduction. Chapters 2 through 4 are the core components of the report. These chapters lay the foundation for the economic analyses of the cost data that are the purpose of this report.

Chapter 2 provides a snapshot of the U.S. construction industry. Historical data on the value of construction put in place are used to highlight the NFPA procedure for estimating the annual cost of fire protection in buildings. Information from the U.S. Department of Energy's Commercial Buildings Energy Consumption Survey[8] is then used to summarize the characteristics of commercial buildings, with special emphasis on buildings over 10 floors. For the most part, these buildings are more than 120 ft (37 m) high.

The methodology and the standardized methods employed in this report to measure the economic performance of selected egress-related requirements are described in Chapter 3. A standardized format for summarizing the results of an economic evaluation is also presented.[9]

Chapter 4 presents information on the five prototypical buildings used to derive cost data for selected egress-related requirements. A key objective of Chapter 4 is analyzing the cost-effectiveness of the required installation of an additional exit stair in buildings over 120 ft (37 m) high versus the alternative of installing an occupant evacuation elevator system. Cost data on exit stairs are first presented along with information on the incremental costs of increasing the width of exit stairs and the implications of installing an additional exit stair for buildings over 120 ft (37 m) high. Cost data on occupant evacuation elevator systems and fire service access elevator systems for the four prototypical buildings over 120 ft (37 m) high are then presented. A cost-effectiveness analysis is then presented where occupant evacuation elevator systems are compared to

---

[8] United States Department of Energy: Energy Information Administration. "Commercial Buildings Energy Consumption Survey." (Washington DC: United States Department of Energy, September 2008).
[9] For a detailed description of the ASTM summary format, see ASTM International. "Standard Guide for Summarizing Economic Impacts of Building Related Projects," E 2204, *Annual Book of ASTM Standards: 2008*, Vol. 04.11. West Conshohocken, PA: ASTM International.

the installation of an additional exit stair for the four prototypical buildings over 120 ft (37 m) high. The results of the cost-effectiveness analysis are then summarized in a standardized format. These results demonstrate that occupant evacuation elevators are a cost-effective alternative to the installation of an additional exit stair in the four prototypical buildings over 120 ft (37 m) high.

Chapter 5 concludes the report with a summary and recommendations for further research.

Appendix A documents the specifications, assumptions, and cost estimating relationships used to develop the costs of the occupant evacuation elevator systems for the four prototypical buildings over 120 ft (37 m) high presented in Chapter 4. The objective of providing the cost data contained in Appendix A is to help other researchers and practitioners to evaluate improved egress system designs that provide for efficient and timely evacuation of occupants, including those with disabilities.

Appendix B documents the specifications, assumptions, and cost estimating relationships used to develop the costs of the fire service access elevator systems for the four prototypical buildings over 120 ft (37 m) high presented in Chapter 4. The objective of providing the cost data contained in Appendix B is to help other researchers and practitioners to evaluate elevator system designs that facilitate more efficient fire department operations.

1.4   Assumptions and Limitations

The specifications, assumptions, and cost estimating relationships of the occupant evacuation elevators were developed in consultation with industry experts. The elevator configuration and related cost elements, described in Appendix A, are representative of one design possibility. Others may exist.

It is important to understand how changes to the assumed cost structure will affect the results. The costs can be separated into two categories: initial and reoccurring costs. Initial costs include those related to water protection, signage, lobby status indicator, two-way communication system, protection of wiring or cables, and the lobby enclosures. Reoccurring costs include those required for maintenance. Changes to the initial costs will affect the first-cost of the occupant evacuation elevators, and are easier to assess their effect on cost-effectiveness. A reduction in assumed initial costs will only increase the cost-effectiveness of occupant evacuation elevators. An increase in the initial costs will have the opposite effect. Only when an increase exceeds the (positive) present value net savings will the occupant evacuation elevators become uneconomic. Changes in the assumed reoccurring costs to economic results are more difficult to (quickly) assess, as a life-cycle analysis is required.

2 Annual Cost of Fire Protection

2.1 Value of Construction Put in Place

In 2008, the latest year for which construction data are available, the construction industry's contribution to gross domestic product (GDP) was $582 billion, or 4.1 % of GDP.[10] In 2008, the value of construction put in place was $1072 billion ($749 billion for new construction, $323 billion for renovation).[11] Maintenance and repair added another $133 billion.[12]

A major investment cost in building construction is fire protection. Fire protection measures include but are not limited to building safety features concerned with extinguishment (e.g., sprinklers), containment (e.g., compartmentation), passive resistance (e.g., fire resistive materials), detection and alarm (e.g., smoke detectors), and egress (e.g., exit stairs). The focus of this report is on egress-related measures in new building construction.

The National Fire Protection Association (NFPA) publishes statistics on the total cost of fire in the U.S.[13] A key component of these statistics is an estimate of the annual costs of fire protection in buildings. The annual costs of fire protection in buildings published by NFPA are based on the Value of Construction Put in Place reported by the U.S. Census Bureau in its C30 report. Table 2.1 reports the values of the key components used to develop the annual estimates of the costs of fire protection in buildings for calendar years 2002 through 2008. Three types of building construction are summarized in the table: Private Residential, Private Nonresidential, and Public Buildings. Specific categories of construction are reported under Private Nonresidential (e.g., Lodging) and Public Buildings (e.g., Office). Private Nonresidential and Public Buildings may be grouped under the more general heading of commercial buildings. Several non-building categories included in the C30 report have been eliminated from Table 2.1.[14]

---

[10] Bureau of Economic Analysis. "Gross-Domestic-Product-(GDP)-by-Industry Data." *Industry Economic Accounts* (Washington, DC: Bureau of Economic Analysis), http://www.bea.gov/bea/dn2/gdpbyind_data.htm (accessed December 2009).

[11] United States Census Bureau: Manufacturing and Construction Division. "Annual Value of Construction Put in Place." *Current Construction Report (CCR) C30 (*Washington, DC: United States Census Bureau, August 1, 2009), http://www.census.gov/const/C30/total.pdf (accessed December 2009).

[12] The value for maintenance and repair is calculated by using the ratio of maintenance and repair to new construction put in place from the 1997 census and multiplying it by the current value for new construction put in place.

[13] Hall, J.R. 2008. "The Total Cost of Fire in the United States." Quincy, MA: National Fire Protection Association.

[14] Private nonresidential construction excludes transportation, communication, power, sewage and waste disposal, and water supply. Public building construction includes all public construction except transportation, power, highway and street, sewage and waste disposal, water supply, and conservation and development.

Table 2.1 Value of Construction Put in Place for Selected Construction Types: 2002 to 2008

	Annual Value of Construction Put in Place						
	(millions of dollars)						
Construction Type	2002	2003	2004	2005	2006	2007	2008
Private Residential	396,696	446,035	532,900	611,899	613,731	493,246	350,078
Private Nonresidential	179,092	173,614	187,570	203,325	235,098	283,046	310,491
Lodging	10,467	9,930	11,982	12,666	17,624	27,481	35,379
Office	35,296	30,579	32,879	37,276	45,680	53,815	57,084
Other	59,008	57,505	63,195	66,584	73,368	85,858	81,495
Health care	22,438	24,217	26,272	28,495	32,016	35,588	39,101
Educational	13,109	13,424	12,701	12,788	13,839	16,691	18,585
Religious	8,335	8,559	8,153	7,715	7,740	7,522	7,097
Public safety	217	185	289	408	419	595	650
Amusement and recreation	7,478	7,781	8,432	7,507	9,326	10,193	10,316
Manufacturing	22,744	21,434	23,667	29,886	35,086	45,303	60,784
Public Building	100,673	100,125	101,349	105,221	112,537	131,366	141,551
Public Housing	5,264	5,216	5,508	5,608	6,083	7,222	7,330
Office	8,982	8,839	9,525	8,487	8,507	11,445	13,222
Other	3,512	4,024	3,862	3,658	3,345	3,827	3,447
Health care	4,701	5,112	5,912	5,935	6,456	8,179	8,598
Educational	60,753	60,892	61,549	66,899	71,089	80,068	85,496
Public safety	7,610	6,976	6,730	6,906	7,350	9,606	12,286
Amusement and recreation	9,851	9,066	8,263	7,728	9,707	11,019	11,172

The annual costs of fire protection in buildings for calendar years 2002 through 2008 are reported in Table 2.2. Table 2.2 reports the estimated cost of fire protection for each of the three types of construction—Private Residential, Private Nonresidential, and Public Buildings—as well as for the overall TOTAL. These values are obtained by multiplying the value of construction put in place by 2.5 % for Private Residential and 4.0 % for Public Buildings construction,[15] and 12.0 % for Private Nonresidential.[16] Applying the NFPA procedure to the most recent data on value of construction put in place, produces an estimated cost of fire protection of $51.7 billion in 2008. The values for each type of construction in 2008 are: $8.8 billion for Private Residential, $37.3 billion for Private Nonresidential, and $5.7 billion for Public Buildings. The bulk of the estimated cost of fire protection, $42.9 billion, is associated with commercial buildings.

---

[15] Apostolou, J.J., D.L. Bowers, and C.M. Sullivan. 1978. "The Nation's Annual Expenditure for the Prevention and Control of Fire." Project Report. Worcester, MA: Worcester Polytechnic Institute.
[16] Meade, W.P. 1991. "A First Pass at Computing Fire Safety in a Modern Society." NIST-GCR-91-592. Gaithersburg, MD: National Institute of Standards and Technology.

Table 2.2 Annual Cost of Fire Protection in Buildings: 2002 to 2008

| | Annual Cost of Fire Protection | | | | | | |
| | (millions of dollars) | | | | | | |
Construction Type	2002	2003	2004	2005	2006	2007	2008
Private Residential	9,917	11,151	13,323	15,297	15,343	12,331	8,752
Private Nonresidential	21,491	20,834	22,508	24,399	28,212	33,966	37,259
Public Building	4,027	4,005	4,054	4,209	4,501	5,255	5,662
TOTAL	35,435	35,990	39,885	43,905	48,057	51,551	51,673

The $51.7 billion estimated cost for fire protection in buildings in 2008 provides a useful reference point vis-à-vis the value of construction put in place. Unfortunately, neither the NFPA procedure nor the source documents upon which the NFPA procedure is based allows us to break out the cost of egress-related measures from the other fire protection measures. Reference documents, such as RS Means Building Construction Cost Data,[17] provide costs per unit of floor area for many types of commercial buildings, including but not limited to: Office, Healthcare, Education, and Lodging. These costs per unit of floor area (e.g., $/ft^2) are often reported in terms of a distribution of values, including: low (25th percentile), median (50th percentile), and high (75th percentile). In addition, costs per unit of floor area for many types of commercial buildings, such as Offices, are often broken out by low-rise, mid-rise, and high-rise. However, a close examination of such reference documents shows considerable variation in cost per unit of floor area as well as in building characteristics. Also, the breakouts for cost per unit of floor area provided in most reference documents are highly aggregated (e.g., they combine mechanical and electrical systems). Therefore, to get at egress-related costs requires an in-depth approach involving analyses of both the characteristics of commercial buildings and cost estimates tied to specific prototypical buildings.

2.2 Characteristics of Commercial Buildings

Characteristics of commercial buildings are analyzed to establish the linkage between the national level cost estimates presented in Table 2.2 and the egress-related cost estimates presented in Chapter 4, Appendix A, and Appendix B of this report. This section focuses on key characteristics involving floor area and year of construction, and concludes with an analysis of buildings with 11 or more floors. Particular emphasis is placed on high-rise Office Buildings, Healthcare Facilities, Educational Facilities, and Lodgings.

Although there are a number of data sets that allow in-depth analyses of the commercial sector, the data associated with the Department of Energy's (DoE's) Commercial Building Energy Consumption Survey (CBECS) is an ideal source for summarizing the characteristics of the commercial sector's stock of buildings. The CBECS collects information on physical characteristics of commercial buildings, building use and

---

[17] Reed Construction Data, Inc. 2008. "RS Means Construction Cost Data." 66th Edition. Kingston, MA: Reed Construction Data, Inc.

occupancy patterns, equipment use, conservation features and practices, and types and uses of energy in buildings. The survey is conducted in two stages, the Building Characteristics Survey and the Energy Suppliers Survey. Our focus is on the Building Characteristics Survey. The types of buildings covered by the CBECS are similar to those appearing in the C30 report, which facilitates a detailed examination of the commercial sector. The most recent DoE CBECS was conducted in 2003; it provides detailed information on the size, age, and other characteristics of commercial buildings. In 2003, there were 4.86 million commercial buildings and 71.66 billion ft^2 (6.66 billion m^2) of commercial floorspace in the U.S. Table 2.3 summarizes the distribution and total floorspace of commercial buildings in 2003 in total and by principal building activity; it is derived from data posted on the CBECS 2003 website.[18]

Table 2.3 Number of Commercial Buildings and Total Floorspace by Principal Building Activity

Principal Building Activity	Number of Buildings in Thousands	Total Floorspace (Millions)	
		ft^2	m^2
All Buildings	*4,859*	*71,658*	*6,657*
Education	386	9,874	917
Food Sales and Service	523	2,909	270
Health Care	129	3,163	294
Lodging	142	5,096	473
Mercantile	657	11,192	1,040
Office	824	12,208	1,134
Public Assembly	277	3,939	366
Public Order and Safety	71	1,090	101
Religious Worship	370	3,754	349
Service	622	4,050	376
Warehouse and Storage	597	10,078	936
Other	261	4,305	400

Public use data files were downloaded from the CBECS website. These "microdata" files enable us to perform more detailed data sorts than the reports posted on the CBECS website. CBECS groups buildings into eight size categories and into eight age categories. The vast majority of commercial buildings were found in the smallest size categories, with more than half in the smallest category and nearly three-quarters in the two smallest categories. Most commercial buildings, once constructed, are expected to last for decades or longer. New buildings are constructed each year and older buildings

---

[18] United States Department of Energy: Energy Information Administration. "Commercial Buildings Energy Consumption Survey." (Washington DC: United States Department of Energy, September 2008), http://www.eia.doe.gov/emeu/cbecs/cbecs2003/detailed_tables_2003/2003set1/2003excel/a1.xls (accessed December 2008).

are demolished, but the commercial building stock at any point in time is dominated by older buildings. More than 50 % of all commercial buildings and total floorspace were constructed prior to 1980, and more than 25 % of buildings and floorspace were constructed prior to 1960.

Figures 2.1 through 2.4 provide detailed snapshots of the nation's stock of commercial buildings. In each figure, information is classified along one of two major dimensions, either by building size, measured in terms of a building's total floorspace, or by building age, measured in terms of a building's year of construction. Each set of figures (e.g., Figures 2.1 and 2.2, and Figures 2.3 and 2.4) uses the same bar chart format to facilitate comparisons of characteristics.

Figure 2.1 records the distribution of the number of commercial buildings by building size. The DoE size categories are specified in customary units; they range from 1001 ft^2 to 5000 ft^2 (93.0 m^2 to 464.5 m^2) for the smallest size category to over 500 000 ft^2 (over 46 451.5 m^2) for the largest size category. Figure 2.1 shows clearly that smaller buildings dominate the key category of commercial buildings. More than half of the stock of commercial buildings (2 586 000 of the 4 859 000) is contained in the smallest size category and almost three-fourths (3 534 000 of the 4 859 000) in the two smallest size categories. By contrast, only five percent or 255 000 buildings are contained in the four largest size categories. However, total floorspace for the four largest size categories is 35.7 billion ft^2 (3.3 billion m^2) or about half the total floorspace across all size categories (i.e., 71.7 billion ft^2 (6.7 billion m^2)).

Figure 2.2 records the distribution of the number of commercial buildings by year of construction. This report uses six "year of construction" categories instead of the eight employed by CBECS. The three "oldest" CBECS year of construction categories are grouped into the 1960 or before year of construction category. All other year of construction categories are the same as in CBECS, The year of construction categories used in this report are 1960 or before, 1960 to 1969, 1970 to 1979, 1980 to 1989, 1990 to 1999, and 2000 to 2003. Figure 2.2 shows that there were more commercial buildings constructed before 1960 (1 442 000 of 4 859 000) than in any other single category, with the next largest number in the 1990 to 1999 year of construction category (917 000 of 4 859 000). Note that the 2000 to 2003 year of construction category contains the smallest number of buildings. This is because the most recent CBECS was conducted in 2003 and thus does not include any buildings constructed since then. The 2007 CBECS, scheduled for release in 2010, will contain a more complete picture of post 2000 commercial building construction.

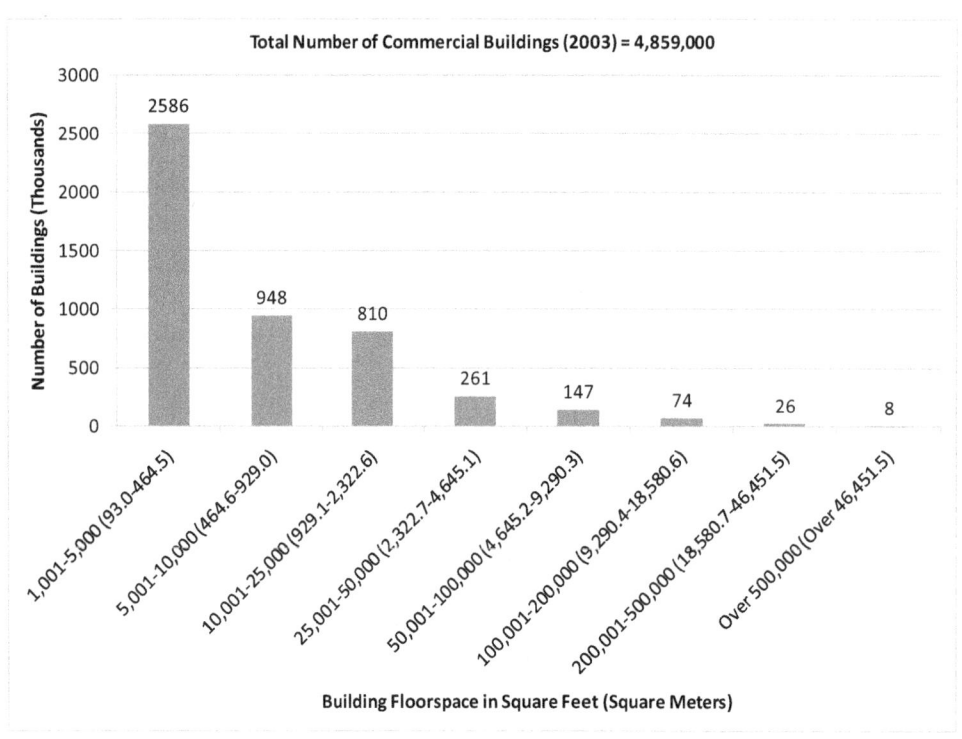

Figure 2.1 Number of Commercial Buildings by Size Category: 2003

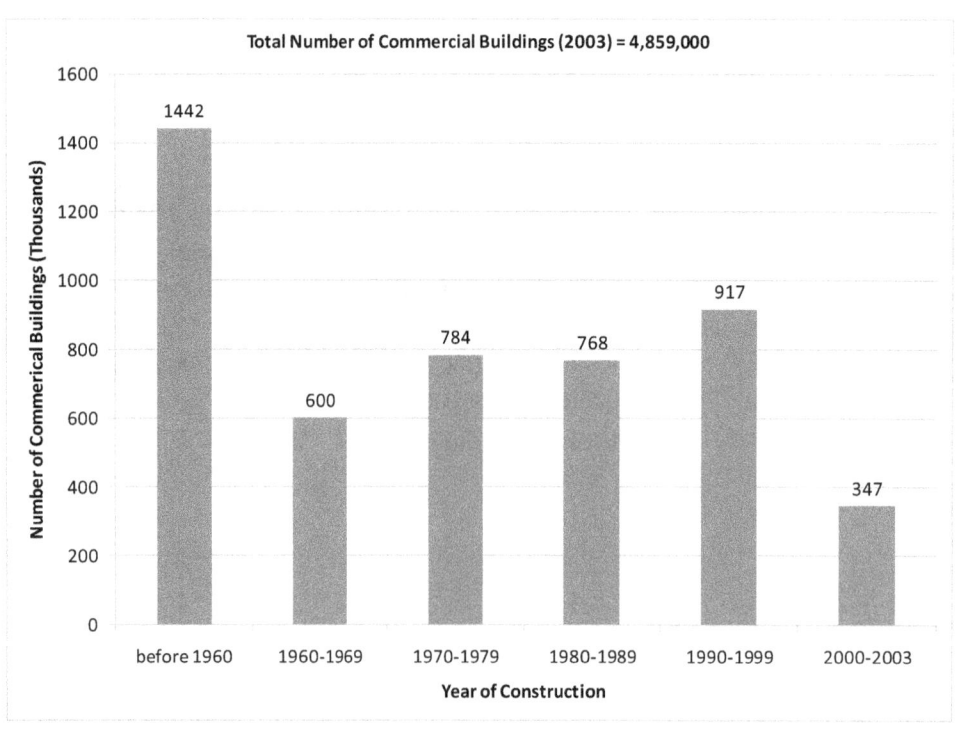

Figure 2.2 Number of Commercial Buildings by Year of Construction: 2003

Figure 2.3 shows the distribution of total floorspace (i.e., nationwide) by building size (i.e., per building). The total floorspace of commercial buildings first increases for the three smallest size categories and then trends downward for the five largest size categories. Figure 2.3 shows that the largest amount of total floorspace is in the third smallest size category (i.e., 10 001 ft^2 to 25 000 ft^2 (929.1 m^2 to 2322.6 m^2)). Although the largest size category (i.e., over 500 000 ft^2 (over 46 451.5 m^2)) contains the smallest number of buildings (refer to Figure 2.1), it contains a significant proportion of total floorspace.

Figure 2.4 introduces an additional characteristic, the number of floors in the building. This characteristic serves to sharpen the distinctions between the buildings in each size category. Figure 2.4 uses the CBECS classification scheme to group commercial buildings into five "number of floor" categories. The number of floors categories are one floor, two floors, three floors, four to ten floors, and 11 or more floors. In Figure 2.4, each number of floors category is coded by shading; a legend is provided on the figure to match the number of floors category to a specific bar in each of the eight size categories. Figure 2.4 shows clearly that commercial buildings with one and two floors dominate the smaller size categories. With the exception of the peak in the third smallest size category (i.e., 10 001 ft^2 to 25 000 ft^2 (929.1 m^2 to 2322.6 m^2)), floorspace for commercial buildings with one floor tends to decrease as the building size gets larger. Figure 2.4 shows that buildings in the three smallest size categories, which contain the largest number of commercial buildings as shown in Figure 2.1, are constructed largely with one and two floors. For the three smallest size categories (i.e., 1001 ft^2 to 25 000 ft^2 (93.0 m^2 to 2322.6 m^2)), the ratio of total floorspace for buildings with one floor to those with two floors is about 2:1. Reference to Figure 2.4 shows that commercial buildings with four to ten floors first increase in total floorspace, reach a peak at 100 001 ft^2 to 200 000 ft^2 (9290.4 m^2 to 18 560.6 m^2), and then decline. Buildings with 11 or more floors increase steadily in total floorspace from the 50 001 ft^2 to 100 000 ft^2 (4645.2 m^2 to 9290.3 m^2) size category to the over 500 000 ft^2 (over 46 451.5 m^2) size category.

Figures 2.5 and 2.6 provide a more detailed look at commercial buildings with 11 or more floors. Both figures are constructed as pie charts. For the purpose of this report, we classify buildings with 11 or more floors as high-rise buildings. Figure 2.5 contains information on the number of high-rise commercial buildings by principal building activity. Reference to the figure reveals that of the 9700 high-rise commercial buildings 57 % are Office Buildings. Lodging (e.g., hotels) account for the second largest proportion at 27 %. Healthcare and Educational Facilities, at 5 % and 0 %, respectively, account for only a small proportion of the total number of high-rise commercial buildings. Note that all Other Commercial buildings only accounts for 11 % of the total.

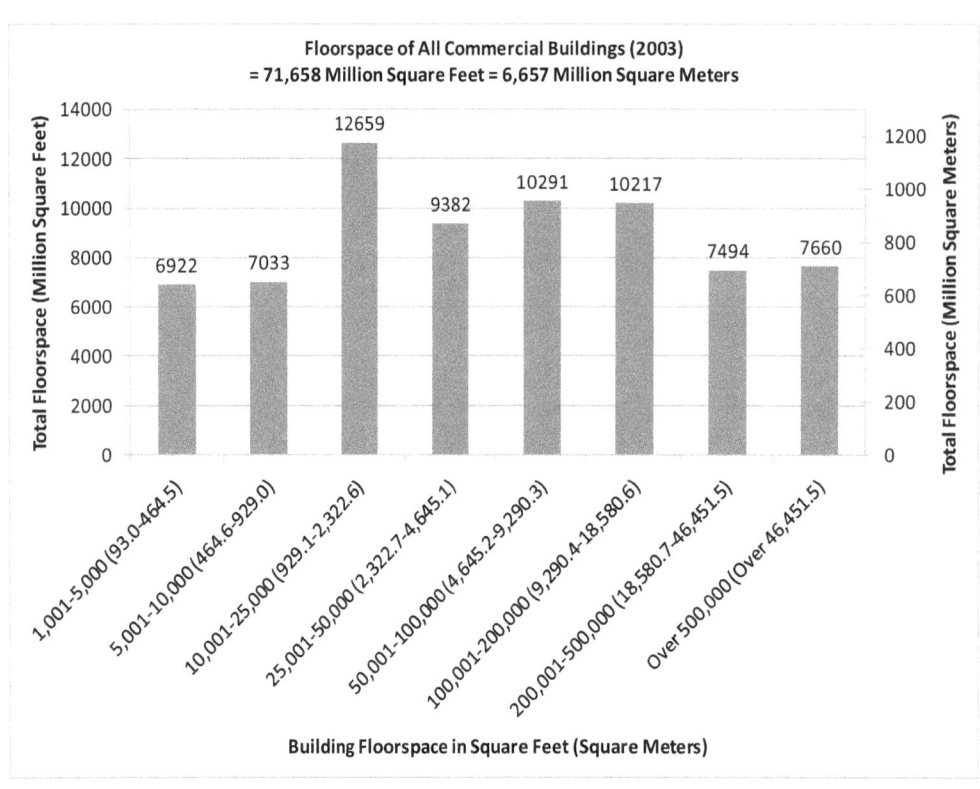

Figure 2.3 Total Floorspace of Commercial Buildings by Size Category: 2003

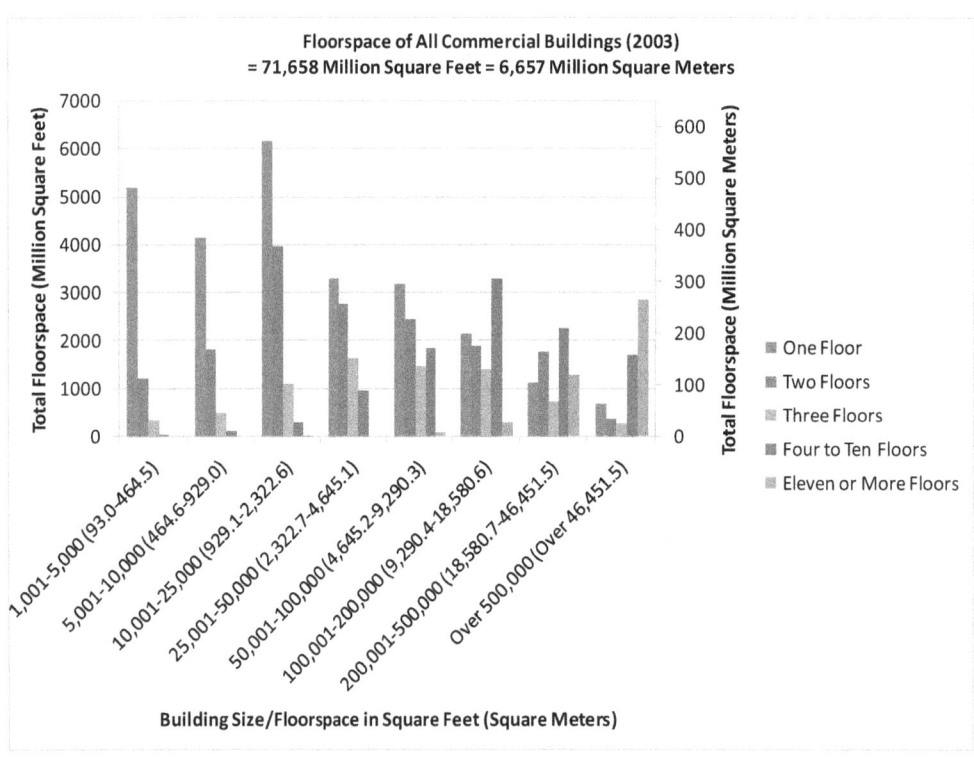

Figure 2.4 Total Floorspace by Size Category and Number of Floors for All Commercial Buildings: 2003

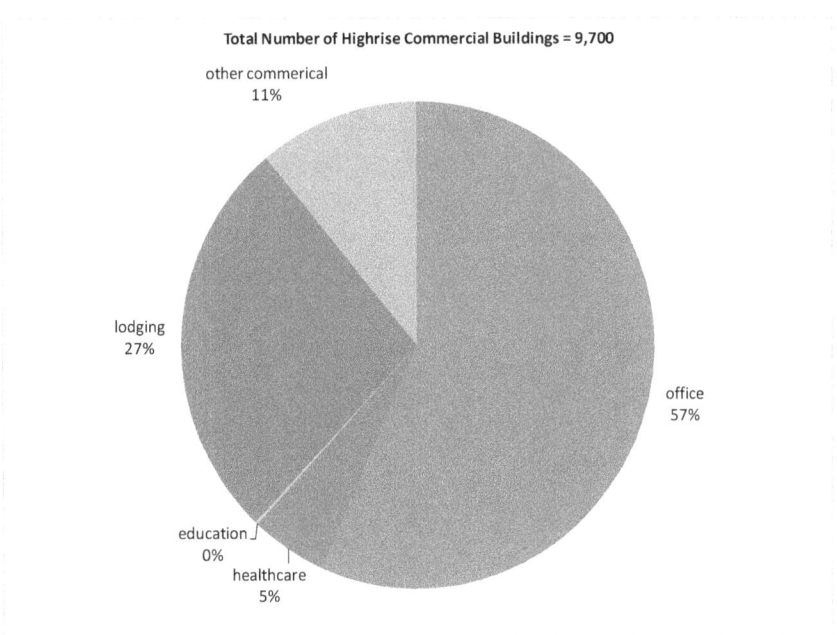

Figure 2.5 Breakdown of High-rise Commercial Buildings by Number[19]

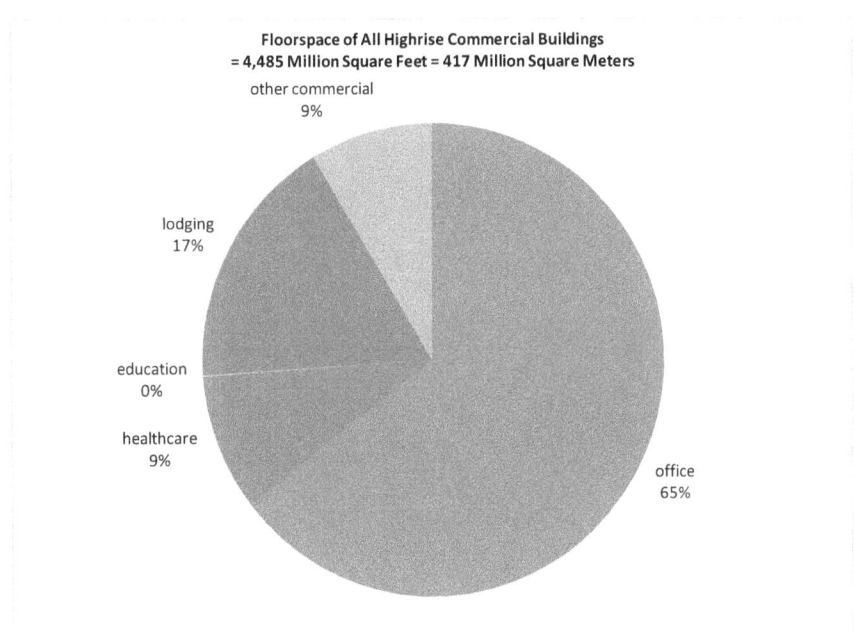

Figure 2.6 Breakdown of High-rise Commercial Buildings by Floorspace[20]

---

[19] Source: Information on all commercial buildings comes from 2003 CBECS.
[20] Source: Information on all commercial buildings comes from 2003 CBECS.

Figure 2.6 contains information on the distribution of floorspace by principal building activity.[21] The figure shows that 4485 million ft^2 (417 million m^2) are in high-rise commercial buildings. Once again, Office Buildings at 65 % account for the largest proportion of total floorspace. Lodging accounts for 17 %. Healthcare and Other Commercial complete the total with 9 % each.

---

[21] Information on high-rise apartment buildings has not been incorporated into Figures 2.5 and 2.6 due to a variety of data classification issues. Readers interested in apartment buildings data are referred to DoE's Residential Energy Consumption Survey (RECS); United States Department of Energy: Energy Information Administration. "Residential Energy Consumption Survey." (Washington DC: United States Department of Energy, April 2008),
http://www.eia.doe.gov/emeu/recs/recs2005/hc2005_tables/detailed_tables2005.html (accessed December 2008).

3 Methodology for Conducting an Economic Analysis of Egress-Related Costs

This chapter focuses on laying out a methodology for conducting and summarizing an economic analysis of egress-related costs. The methodology is based on two types of analysis, four measures of economic performance, and a format for summarizing the results of an economic analysis of egress-related costs. The two types of analysis are baseline analysis and sensitivity analysis. They are described in Section 3.1. The four measures of economic performance are life-cycle cost, present value of net savings, savings-to-investment ratio, and adjusted internal rate of return. They are described in Section 3.2. The format for summarizing the results of an economic analysis of egress-related costs is described in Section 3.3.

As noted in Chapter 1, there are four egress-related requirements for which cost data are presented in this report. Two of the four requirements are "alternatives" in the context of the material presented in this chapter and elsewhere in this report. The two alternatives are: (1) the required installation of an additional exit stair in buildings over 420 ft (128 m) high or (2) the installation of an occupant evacuation elevator system. The alternatives are evaluated for all buildings exceeding 120 ft (37 m) in height.

3.1 Types of Analysis

An economic analysis of egress-related costs is a complicated process, entailing two distinct levels of analysis. This "analysis strategy" systematically adds increased detail to the decision-making process. The first level is referred to as the baseline analysis. Here we are working with our best-guess estimates (often some values are not known with certainty). The baseline analysis provides a frame of reference for the treatment of uncertainty, which is the focus of the second level—sensitivity analysis—which systematically varies selected sets of data elements to measure their economic impacts on project outcomes.

3.1.1 Baseline Analysis

The starting point for conducting an economic evaluation is to do a baseline analysis. In the baseline analysis, all data elements entering into the calculations are fixed. For some data, the input values are considered to be known with certainty. Other data are considered uncertain and their values are based on some measure of central tendency, such as the mean or the median, or input from subject matter experts. Baseline data represent a fixed state of analysis. For this reason, the analysis results are referred to as the baseline analysis. The term baseline analysis is used to denote a complete analysis in all respects but one; it does not address the effects of uncertainty.

3.1.2 Sensitivity Analysis

Sensitivity analysis measures the impact on project outcomes of changing the values of one or more key data elements about which there is uncertainty. Sensitivity analysis can be performed for any measure of economic performance (e.g., life-cycle cost or present value of net savings). Therefore, a sensitivity analysis complements the baseline analysis by evaluating the changes in output measures when selected data inputs are allowed to vary about their baseline values. There are two basic types of sensitivity analyses: deterministic and probabilistic.

The key advantage of deterministic sensitivity analyses is that they are easily constructed and computed and the results are easy to explain and understand. Their disadvantage is that they do not produce results that can be tied to probabilistic levels of significance.

Monte Carlo simulation, a widely used probabilistic sensitivity analysis method, varies a small set of key parameters either singly or in combination according to an experimental design. Associated with each key parameter is a probability distribution function from which values are randomly sampled. The major advantage of the Monte Carlo simulation technique is that it permits the effects of uncertainty to be rigorously analyzed through reference to a derived distribution of project outcome values. Their disadvantage is that data on the probability distribution are generally not well known.

3.2 Overview of Evaluation Methods

Numerous methods are available for measuring the economic performance of investments in buildings and building systems. Use ASTM Standard Guide E 1185[22] to identify types of decisions on building designs and systems that require economic evaluation and to match the technically appropriate economic methods with those decisions.

Four economic evaluation methods addressed in ASTM Standard Guide E 1185 apply to the economic analysis of egress-related costs: (1) life-cycle costs (ASTM Standard Practice E 917[23]); (2) present value net savings (ASTM Standard Practice E 1074[24]); (3)

---

[22] ASTM International. "Standard Guide for Selecting Economic Methods for Evaluating Investments in Buildings and Building Systems," E 1185, *Annual Book of ASTM Standards: 2008*, Vol. 04.11. West Conshohocken, PA: ASTM International.

[23] For a detailed description of the ASTM life-cycle cost standard, see ASTM International. "Standard Practice for Measuring Life-Cycle Costs of Buildings and Building Systems," E 917, *Annual Book of ASTM Standards: 2008*, Vol. 04.11. West Conshohocken, PA: ASTM International.

[24] For a detailed description of the ASTM present value of net savings standard, see ASTM International. "Standard Practice for Measuring Net Benefits and Net Savings for Investments in Buildings and Building Systems," E 1074, *Annual Book of ASTM Standards: 2008*, Vol. 04.11. West Conshohocken, PA: ASTM International.

savings-to-investment ratio (ASTM Standard Practice E 964[25]); and (4) adjusted internal rate of return (ASTM Standard Practice E 1057[26]).

More than one method can be technically appropriate for many design and system decisions. If more than one method is technically appropriate, use all that apply, since many decision makers need information both on measures of magnitude (life-cycle costs and present value net savings) and of return (savings-to-investment ratio and adjusted internal rate of return) to assess economic performance.

3.2.1 Life-Cycle Cost Method

The life-cycle cost (LCC) method measures, in present-value or annual-value terms, the sum of all relevant costs associated with owning and operating a constructed facility over a specified period of time. The basic premise of the LCC method is that to an investor or decision maker all costs arising from that investment decision are potentially important to that decision, including future as well as present costs associated with egress-related requirements. Applied to constructed facilities, the LCC method encompasses all relevant costs over a designated study period, including the costs of designing, purchasing/leasing, constructing/installing, operating, maintaining, repairing, replacing, and disposing of a particular design or system. Should any pure benefits result (e.g., increased rental income due to improvements), include them in the calculation of LCC.

The LCC method is particularly suitable for determining whether the higher initial cost of a constructed facility or system specification is economically justified by lower future costs when compared to an alternative with a lower initial cost but higher future costs. If a design or system specification has both a lower initial cost and lower future costs relative to an alternative, an LCC analysis is not needed to show that the former is economically preferable.

Denote the alternative with the lowest initial investment cost (i.e., first cost) as the base case. The LCC method compares alternative, mutually exclusive, designs or systems that satisfy a given functional requirement—in this case, two of the four egress-related requirements specified in the IBC—on the basis of their life-cycle costs to determine which is the least-cost means (i.e., minimizes life-cycle cost) of satisfying that requirement over a specified study period. With respect to the base case, an alternative is economically preferred if, and only if, it results in lower life-cycle costs.

Again, the two alternatives are: (1) the required installation of an additional exit stair in buildings over 420 ft (128 m) high or (2) the installation of an occupant evacuation

---

[25] For a detailed description of the ASTM savings-to-investment ratio standard, see ASTM International. "Standard Practice for Measuring Benefit-to-Cost and Savings-to-Investment Ratios for Investments in Buildings and Building Systems," E 964, *Annual Book of ASTM Standards: 2008*, Vol. 04.11. West Conshohocken, PA: ASTM International.

[26] For a detailed description of the ASTM adjusted internal rate of return standard, see ASTM International. "Standard Practice for Measuring Internal Rate of Return and Adjusted Internal Rate of Return for Investments in Buildings and Building Systems," E 1057, *Annual Book of ASTM Standards: 2008*, Vol. 04.11. West Conshohocken, PA: ASTM International.

elevator system, although these are evaluated for all buildings exceeding 120 ft (37 m) in height.

The alternative—installation of an additional exit stair (base case) or installation of an occupant evacuation elevator system—that results in the lowest life-cycle cost is designated as the most cost-effective.

3.2.2 Present Value Net Savings

The present value of net savings (PVNS) method is reliable, straightforward, and widely applicable for finding the economically efficient choice among investment alternatives. It measures the net savings from investing in a given alternative instead of investing in the foregone opportunity (e.g., some other alternative or the base case).
The PVNS for a given alternative, vis-à-vis the base case, equals their difference in life-cycle costs. Any pure benefits that result (e.g., increased rental income due to improvements) are included in the calculation of PVNS, since they are included in the LCC calculation.

With respect to the base case, if PVNS is positive for a given alternative the investment is economic; if it is zero, the investment is as good as the base case; if it is negative, the investment is uneconomical.

The installation of an occupant evacuation elevator system vis-à-vis the exit stair base case is cost-effective if the PVNS is greater than zero.

3.2.3 Savings-to-Investment Ratio

The savings-to-investment ratio (SIR) is a numerical ratio whose value indicates the economic performance of a given alternative instead of investing in the foregone opportunity. The SIR is savings divided by investment costs. The LCC method provides all of the necessary information to calculate the SIR. The SIR for a given alternative is calculated vis-à-vis the base case.

The numerator equals the difference in the present value of non-investment costs between the base case and the given alternative. The denominator equals the difference in the present value of investment costs for the given alternative and the base case. A ratio less than 1.0 indicates that the given alternative is an uneconomic investment relative to the base case; a ratio of 1.0 indicates an investment whose benefits or savings just equal its costs; and a ratio greater than 1.0 indicates an economic project.

The installation of an occupant evacuation elevator system vis-à-vis the exit stair base case is cost-effective if the SIR is greater than 1.0.

3.2.4 Adjusted Internal Rate of Return

The adjusted internal rate of return (AIRR) is the average annual yield from a project over the study period, taking into account reinvestment of interim receipts. The reinvestment rate in the AIRR calculation is equal to the minimum acceptable rate of return (MARR), which is assumed to equal the discount rate. When the reinvestment rate is made explicit, all investment costs are easily expressible as a time equivalent initial outlay (i.e., a value at the beginning of the study period) and all non-investment cash flows as a time equivalent terminal amount. This allows a straightforward comparison of the amount of money that comes out of the investment (i.e., the terminal value) with the amount of money put into the investment (i.e., the time equivalent initial outlay).

The AIRR is defined as the interest rate applied to the terminal value, which equates (i.e., discounts) it to the time equivalent value of the initial outlay of investment costs. It is important to note that all investment costs are discounted to a time equivalent initial outlay using the discount rate.

With regard to the base case, if the AIRR is greater than the discount rate (also referred to as the hurdle rate), then investment in the given alternative is economic; if the AIRR equals the discount rate, the investment is as good as the base case; if AIRR is less than the discount rate, the investment is uneconomical.

The installation of an occupant evacuation elevator system vis-à-vis the exit stair base case is cost-effective if the AIRR is greater than the discount rate.

3.3 Presentation and Analysis of the Results of an Economic Analysis

The presentation and analysis of the results of an economic analysis are central to understanding and accepting its findings. If the presentation is clear and concise, and if the analysis strategy is logical, complete, and carefully spelled out, then the results will stand up under close scrutiny. The purpose of this section is to outline a generic framework for economic analyses that meets the two previously cited conditions. The generic framework is built upon the following three factors: (1) the significance of the study effort; (2) the analysis strategy; and (3) the calculation of key benefit and cost measures.[27] A specific framework, tailored to BFRL, is given in Exhibit 3.1; it is also used as the basis for summarizing the economic analysis of egress-related costs (see Section 4.4.3).

The discussion that follows relates the three factors for the generic framework referenced above to the specific framework given in Exhibit 3.1. Exposition of the generic framework serves two purposes. First, it provides a means for organizing the way to present material associated with an in-depth economic analysis of egress-related costs. Second, it provides a vehicle for clearly and concisely presenting the salient results of the

---

[27] This framework is based on ASTM Standard Guide E 2204 (ASTM International. "Standard Guide for Summarizing Economic Impacts of Building Related Projects," E 2204, *Annual Book of ASTM Standards: 2008*, Vol. 04.12. West Conshohocken, PA: ASTM International.).

analysis. Such a short summary is appropriate for use by senior managers (e.g., research directors, facilities executives, development executives) as the basis for statements on the benefits of the study.

3.3.1 Significance of Study Effort

This section of an economic analysis sets the stage for the results that follow. The goal at this point is to clearly describe:

(1) why the study is important and how the organization conducting it became involved; and

(2) why some or all of the changes brought about were due to the study organization's contribution.

Emphasis is placed on providing dollar estimates to define the magnitude of the problem. If any non-financial characteristics are of key importance to senior management, list and describe them briefly.[28] A clear tie into the study organization's mission or vision is included to demonstrate why the organization conducting the study is well qualified and well positioned to participate in the study. The section concludes with a statement of the study organization's contribution.

3.3.2 Analysis Strategy

This section of an economic analysis focuses on documenting the steps taken to ensure that the analysis strategy is logical and complete. Particular emphasis is placed on summarizing the key assumptions, including any constraints that limited the scope of the analysis. Responses are provided for key assumptions concerning: (a) the base year for the analysis; (b) the length of the study period; and (c) the discount rate or minimum acceptable rate of return used.

Special emphasis is placed on documenting the *sources and validity* of any data used to make estimates or projections of key benefit and cost measures. This section establishes an audit trail from the raw data, through data manipulations (e.g., represented by equations and formulae), to the results which describe how to determine:

(1) the present value of **total benefits (savings)**;

(2) the present value of **total costs**;

(3) the present value of **net benefits (savings)**; and

(4) the way in which any **additional measures** were calculated.

Decision makers typically experience uncertainty about the correct values to use in establishing basic assumptions and in estimating both first costs and future costs. Specify the assumptions or costs that have a high degree of uncertainty and are likely to have a

---

[28] Refer to ASTM Standard Practice E 1765 (ASTM International. "Standard Practice for Applying Analytical Hierarchy Process (AHP) to Multiattribute Decision Analysis of Investments Related to Buildings and Building Systems," E 1765, *Annual Book of ASTM Standards: 2008,* Vol. 04.12. West Conshohocken, PA: ASTM International.) and its adjunct for guidance on how to present unquantified effects.

significant impact on the results of the analysis. Document the sensitivity of the results to these assumptions or data. ASTM Standard Guide E 1369 recommends techniques for treating uncertainty in parameter values in an economic analysis.[29] It also recommends techniques for evaluating the risk that a project will have a less favorable economic outcome than what is desired or expected. ASTM Standard Practice E 1946 establishes a procedure for measuring cost risk for buildings and building systems, using the Monte Carlo simulation technique as described in ASTM Standard Guide E 1369.[30]

---

[29] ASTM International. "Standard Guide for Selecting Techniques for Treating Uncertainty and Risk in the Economic Evaluation of Buildings and Building Systems," E 1369, *Annual Book of ASTM Standards: 2008,* Vol. 04.11. West Conshohocken, PA: ASTM International.

[30] ASTM International. "Standard Practice for Measuring Cost Risk of Buildings and Building Systems," E 1946, *Annual Book of ASTM Standards: 2008,* Vol. 04.12. West Conshohocken, PA: ASTM International.

Exhibit 3.1 Format for Summarizing the Results of an Economic Analysis

1.a Significance of Study Effort: Describe why the study is important and how BFRL became involved. *Describe the changes brought about by BFRL.*	1.b Key Points: *Highlight two or three key points which convey why this study effort is important.*
2. Analysis Strategy: *Describe how the present value of **total benefits (savings)** was determined.* *Describe how the present value of **total costs** was determined.* *Describe how the present value of **net benefits (savings)** was determined.* *Describe how any **additional measures** were calculated.* *Summarize key data and assumptions: (a) Base year; (b) Length of study period; (c) Discount rate or minimum acceptable rate of return; (d) Data; and (e) other.*	
3.a Calculation of Benefits, Costs, and Additional Measures: **Total Benefits (Savings):** Report the present value of the total benefits (savings). **Total Costs:** Report the present value of the total costs. **Net Benefits (Savings):** Report the present value of net benefits (savings). **Additional Measures:** Report the values of any additional measures calculated.	3.b Key Measures: Report the calculated value of the *Present Value of Net Benefits (PVNB)* **or** *the Present Value of Net Savings (PVNS)* **and** at least one of the following: Benefit-to-Cost Ratio (BCR) *or* Savings-to-Investment Ratio (SIR) Adjusted Internal Rate of Return (AIRR)
	3.c Traceability Cite references to specific ASTM standard practices, ASTM adjuncts, or any other standards, codes, or regulations used.

3.3.3 Calculation of Benefits, Costs, and Additional Measures

This section of an economic analysis focuses on reporting the calculated values of the key benefit and cost measures, as well as any additional measures that are deemed appropriate, and establishing traceability to standardized practices or, where appropriate, to statutory documents or procedures. It consists of three subsections, designated as 3.a, 3.b, and 3.c. Subsection 3.a includes descriptive information as well as calculated values. Subsection 3.b reports calculated values for key measures of economic performance. Subsection 3.c is included to ensure traceability to appropriate national standards, codes, or regulations.

In subsection 3.a, report summaries (e.g., using text, mathematical expressions, tables, graphs, comparative statistics) of the following information:
- (1) the present value of the total benefits (savings);
- (2) the present value of the total costs;
- (3) the present value of net benefits (savings); and
- (4) the values of any additional measures calculated.

In subsection 3.b, report the calculated value of the present value of net benefits *or the present value of net savings* and at least one of the following:
- (a) the benefit-to-cost ratio *or the savings-to-investment ratio*; or
- (b) the adjusted internal rate of return.

In subsection 3.c, cite references to specific ASTM standard practices, ASTM adjuncts, or any other standards, codes, or regulations used.

4 Tabulation and Analysis of Egress-Related Cost Data

4.1 Use of Prototypical Building Designs

Egress-related measures are a major component of any fire protection strategy in buildings. This report tabulates cost data for selected egress-related requirements in five prototypical buildings. The five prototypical buildings range in height from a five-floor, mid-rise building to a 75-floor, high-rise building. Characteristics of the five prototypical buildings are summarized in Table 4.1. Egress-related cost data were compiled from a number of sources, including industry experts, design professionals, and cost estimating guidebooks and software. Cost data are tabulated in a format that facilitates life-cycle cost analyses of selected egress-related requirements. Incremental costs are also tabulated to help assess the implications of changing one or more design parameters.

Table 4.1 Summary Information on the Prototypical Buildings Used in Developing Egress-Related Cost Data

Building	Number of Floors	Building Height	Per Floor Area	Total Floorspace	Cost	
		Feet (ft)	ft^2	ft^2	$ million	$/ft^2$
1	5	60	20,000	100,000	10.0	100.0
2	13	156	25,000	325,000	42.3	130.2
3	28	336	30,000	840,000	147.0	175.0
4	42	504	40,000	1,680,000	504.0	300.0
5	75	900	45,000	3,375,000	1,215.0	360.0

An objective of this chapter is analyzing the cost-effectiveness of the required installation of an additional exit stair in buildings over 120 ft (37 m) high versus the alternative of installing an occupant evacuation elevator system. Although the life-cycle costs of either an additional exit stair or an occupant evacuation elevator system are likely to amount to several million dollars for both Buildings 4 and 5, reference to Table 4.1 demonstrates they represent only a small fraction (i.e., about 0.6 % to 1.2 %) of the construction cost of those buildings.

4.2 Exit Stairs

Much of the cost data presented in this section was provided by William Hunt, Chief Estimator, U.S. General Services Administration (GSA). To distinguish these cost data from other sources, we reference the source as Exit Stair Cost Analysis document.[31] Most of the design guidance for the five prototypical buildings summarized in Table 4.1 was contained in the Exit Stair Cost Analysis document. The section concludes with a calculation of the life-cycle costs of installing and maintaining an additional exit stair in each of the five prototypical buildings, with particular emphasis on prototypical buildings greater than 120 ft (128 m) in height.

---

[31] United States General Services Administration. "Exit Stair Cost Analysis." (Washington, DC: United States General Services Administration, July 12, 2007).

4.2.1 Alternative Exit Stair Configurations

As noted earlier, many of the cost estimates are based on "initial capital cost" data contained in the Exit Stair Cost Analysis document. Since the cost estimates contained in the Exit Stair Cost Analysis document are for July 2007 in Washington, DC, all other cost information presented in this report uses July 2007 as its reference date and the greater Washington, DC metropolitan area as the building's location. The Exit Stair Cost Analysis document data as well as data compiled by the project team are summarized in Tables 4.2 through 4.6. Table 4.2 summarizes cost data for Building 1. Table 4.3 summarizes cost data for Building 2. Table 4.4 summarizes cost data for Building 3. Table 4.5 summarizes cost data for Building 4. Table 4.6 summarizes cost data for Building 5. The cost data summarized in Tables 4.2 through 4.6 are "initial capital costs." The implications of future costs are examined at the end of this section.

Each table is divided into three parts. Part 1 covers the case where there are two exit stairs present. Part 2 covers the case where there are three exit stairs present. Part 3 covers the case where there are four exit stairs present. Each of the three parts is subdivided into three subparts. Subpart 1 covers the case where the nominal width of the exit stair is 44 in (112 cm), the current minimum. Subpart 2 covers the case where the nominal width of the exit stair is 56 in (142 cm), a width based on previous egress studies.[32] Subpart 3 covers the case where the nominal width of the exit stair is 66 in (168 cm). Subpart 3 addresses the recently adopted 2009 International Building Code requirement for an increase of 50 % in the width of exit stairs in new sprinklered buildings for buildings with floor areas exceeding 15 000 ft^2 (1394 m^2).

Four types of cost data are presented in Tables 4.2 through 4.6: total cost of the system as configured (e.g., two exit stairs with a nominal width of 44 in (112 cm)); the cost per unit of floor area; the cost per unit of building height; and the cost per floor. Entries in Tables 4.2 through 4.6 that are tied directly to the Exit Stair Cost Analysis document are shown in **boldface** font.

Specific cost items reported in Tables 4.2 through 4.6 are: the cost of the exit stair system as configured;[33] the cost of installing photoluminous exit path markings (PLEPM) in the exit stair system as configured; and the total cost of the exit stair system and PLEPM. Installation of PLEPM is accomplished by placing one in (2.54 cm) wide marking stripes on the horizontal leading edge of each step, the horizontal leading edge of each landing, and the top surface of each handrail.

---

[32] E.g., see Templer, J.A. Stair Shape and Human Movement, Ph.D. dissertation. New York, NY: Columbia University, (1974).

[33] Individual cost items included in the exit stair assembly are: foundation/slab; steel frame; stair treads; stair landings; stair rails; wall hand rails; exterior wall; exterior doors; roofing; interior partition; interior doors; interior wall finish; interior floor finish; stair pressurization; stair lighting; and exit lights.

For cases where there was not a direct match between the cost data contained in the Exit Stair Cost Analysis document and the system as configured in Tables 4.2 through 4.6, it was necessary to develop a cost estimating relationship to fill the gap. These cost estimating relationships involved both linear interpolations of data contained in the Exit Stair Cost Analysis document and using cost factor models to extrapolate for those cases outside the ranges contained in the Exit Stair Cost Analysis document (e.g., exit stairs with a nominal width of 66 in (168 cm)).

Reference to Table 4.2, where the cost estimates for Building 1 are presented, shows that eight of the 27 Total Cost entries are traceable directly to the Exit Stair Cost Analysis document (i.e., they appear in **boldface** font). This pattern is repeated in Tables 4.2 through 4.6.

Returning to Table 4.2, we see in Part 1.1 that the estimated total cost of the system as configured (i.e., two exit stairs with a nominal width of 44 in (112 cm) and PLEPM installed) is $252 455. This estimate is directly traceable to the Exit Stair Cost Analysis document. Turning to Table 4.2, Part 1.2, we see that the Total Cost estimate of $283 693 is a combination of cost data from the Exit Stair Cost Analysis document and cost estimating relationships developed by the project team. In this case, the cost estimate for the two exit stairs with a nominal width of 56 in (142 cm) is directly traceable to the Exit Stair Cost Analysis document, but the estimate for the PLEPM installation is based on the project team's cost estimating relationship. In this case, the estimated cost of installing PLEPM was based on an interpolation from the 44 in (112 cm) and 57 in (145 cm) values reported in the Exit Stair Cost Analysis document.[34] Turning next to Table 4.2, Part 1.3, we see that expanding the nominal width of the exit stair from 44 in (112 cm) to 66 in (168 cm) increases the Total Cost to $309 724.

Reference to Part 1.1 of Tables 4.2 through 4.6 (two exit stairs each having a nominal width of 44 in (112 cm)) reveals that the cost of exit stairs per unit of floor area declines from $2.50/ft^2 ($26.91/m^2) for Building 1 (five floor mid-rise) to $1.08/ft^2 ($11.63/m^2) for Building 5 (75 floor high-rise). This decline in the cost of exit stairs per unit of floor area is due to the increasing floor "footprint." Reference to Table 4.1 provides the necessary insight. For Building 1, the floor "footprint" is 20 000 ft^2 (1858 m^2). The floor "footprint" increases steadily for Buildings 2, 3, and 4, and reaches a maximum of 45 000 ft^2 (4181 m^2) for Building 5. Note that the number of exit stairs in Part 1.1 of Tables 4.2 through 4.6 is held constant at two, whereas the total floorspace increases from 100 000 ft^2 (9290 m^2) for Building 1 to 3 375 000 ft^2 (313 545 m^2) for Building 5.

---

[34] Although the cost of installing exit stairs with a nominal width of 56 in (142 cm) was included in the Exit Stair Cost Analysis document, no estimates of the cost of installing PLEPM was included for that stair configuration, so they were based on a linear interpolation between the 44 in (112 cm) and 57 in (145 cm) values.

Table 4.2 Exit Stair-Related Cost Data for Building 1: 5 Floors, Height of 60 ft (18 m), and Total Floorspace of 100 000 ft^2 (9290 m^2)

	Total Cost	$/ft^2	$/VLF	$/floor
Cost of Stairs:	**249,772**	2.50	4,162.87	49,954.40
Cost of Photoluminous Exit Markings:	**2,683**	0.03	44.72	536.60
Total Cost:	252,455	2.52	4,207.58	50,491.00

Part 1.2: 2 exit stairs each having a nominal width dimension of 56 inches				
	Total Cost	$/ft^2	$/VLF	$/floor
Cost of Stairs:	**280,278**	2.80	4,671.30	56,055.60
Cost of Photoluminous Exit Markings:	3,415	0.03	56.91	682.95
Total Cost:	283,693	2.84	4,728.21	56,738.55

Part 1.3: 2 exit stairs each having a nominal width dimension of 66 inches				
	Total Cost	$/ft^2	$/VLF	$/floor
Cost of Stairs:	305,700	3.06	5,094.99	61,139.93
Cost of Photoluminous Exit Markings:	4,025	0.04	67.08	804.90
Total Cost:	309,724	3.10	5,162.07	61,944.83

Part 2.1: 3 exit stairs each having a nominal width dimension of 44 inches				
	Total Cost	$/ft^2	$/VLF	$/floor
Cost of Stairs:	**374,658**	3.75	6,244.30	74,931.60
Cost of Photoluminous Exit Markings:	4,025	0.04	67.08	804.90
Total Cost:	**378,683**	3.79	6,311.38	75,736.50

Part 2.2: 3 exit stairs each having a nominal width dimension of 56 inches				
	Total Cost	$/ft^2	$/VLF	$/floor
Cost of Stairs:	420,417	4.20	7,006.95	84,083.40
Cost of Photoluminous Exit Markings:	5,122	0.05	85.37	1,024.42
Total Cost:	425,539	4.26	7,092.32	85,107.82

Part 2.3: 3 exit stairs each having a nominal width dimension of 66 inches				
	Total Cost	$/ft^2	$/VLF	$/floor
Cost of Stairs:	458,550	4.59	7,642.49	91,709.90
Cost of Photoluminous Exit Markings:	6,037	0.06	100.61	1,207.35
Total Cost:	464,586	4.65	7,743.10	92,917.25

Part 3.1: 4 exit stairs each having a nominal width dimension of 44 inches				
	Total Cost	$/ft^2	$/VLF	$/floor
Cost of Stairs:	**499,544**	5.00	8,325.73	99,908.80
Cost of Photoluminous Exit Markings:	5,366	0.05	89.43	1,073.20
Total Cost:	504,910	5.05	8,415.17	100,982.00

Part 3.2: 4 exit stairs each having a nominal width dimension of 56 inches				
	Total Cost	$/ft^2	$/VLF	$/floor
Cost of Stairs:	560,556	5.61	9,342.60	112,111.20
Cost of Photoluminous Exit Markings:	6,829	0.07	113.82	1,365.89
Total Cost:	567,385	5.67	9,456.42	113,477.09

Part 3.3: 4 exit stairs each having a nominal width dimension of 66 inches				
	Total Cost	$/ft^2	$/VLF	$/floor
Cost of Stairs:	611,399	6.11	10,189.99	122,279.87
Cost of Photoluminous Exit Markings:	8,049	0.08	134.15	1,609.80
Total Cost:	619,448	6.19	10,324.14	123,889.67

Values in bold came from the "Exit Stair Cost Analysis" document.

Table 4.3 Exit Stair-Related Cost Data for Building 2: 13 Floors, Height of 156 ft (48 m), and Total Floorspace of 325 000 ft^2 (30 193 m^2)

Part 1.1: 2 exit stairs each having a nominal width dimension of 44 inches				
	Total Cost	$/ft^2	$/VLF	$/floor
Cost of Stairs:	**637,889**	1.96	4,089.03	49,068.38
Cost of Photoluminous Exit Markings:	**8,048**	0.02	51.59	619.08
Total Cost:	**645,937**	1.99	4,140.62	49,687.46

Part 1.2: 2 exit stairs each having a nominal width dimension of 56 inches				
	Total Cost	$/ft^2	$/VLF	$/floor
Cost of Stairs:	**721,490**	2.22	4,624.94	55,499.23
Cost of Photoluminous Exit Markings:	10,243	0.03	65.66	787.92
Total Cost:	731,733	2.25	4,690.60	56,287.15

Part 1.3: 2 exit stairs each having a nominal width dimension of 66 inches				
	Total Cost	$/ft^2	$/VLF	$/floor
Cost of Stairs:	791,158	2.43	5,071.52	60,858.27
Cost of Photoluminous Exit Markings:	12,072	0.04	77.38	928.62
Total Cost:	803,230	2.47	5,148.91	61,786.88

Part 2.1: 3 exit stairs each having a nominal width dimension of 44 inches				
	Total Cost	$/ft^2	$/VLF	$/floor
Cost of Stairs:	**956,834**	2.94	6,133.55	73,602.58
Cost of Photoluminous Exit Markings:	12,072	0.04	77.38	928.62
Total Cost:	**968,906**	2.98	6,210.93	74,531.19

Part 2.2: 3 exit stairs each having a nominal width dimension of 56 inches				
	Total Cost	$/ft^2	$/VLF	$/floor
Cost of Stairs:	1,082,235	3.33	6,937.40	83,248.85
Cost of Photoluminous Exit Markings:	15,364	0.05	98.49	1,181.87
Total Cost:	1,097,599	3.38	7,035.89	84,430.72

Part 2.3: 3 exit stairs each having a nominal width dimension of 66 inches				
	Total Cost	$/ft^2	$/VLF	$/floor
Cost of Stairs:	1,186,736	3.65	7,607.28	91,287.40
Cost of Photoluminous Exit Markings:	18,108	0.06	116.08	1,392.92
Total Cost:	1,204,844	3.71	7,723.36	92,680.33

Part 3.1: 4 exit stairs each having a nominal width dimension of 44 inches				
	Total Cost	$/ft^2	$/VLF	$/floor
Cost of Stairs:	**1,275,778**	3.93	8,178.06	98,136.77
Cost of Photoluminous Exit Markings:	16,096	0.05	103.18	1,238.15
Total Cost:	1,291,874	3.97	8,281.24	99,374.92

Part 3.2: 4 exit stairs each having a nominal width dimension of 56 inches				
	Total Cost	$/ft^2	$/VLF	$/floor
Cost of Stairs:	1,442,980	4.44	9,249.87	110,998.46
Cost of Photoluminous Exit Markings:	20,486	0.06	131.32	1,575.83
Total Cost:	1,463,466	4.50	9,381.19	112,574.29

Part 3.3: 4 exit stairs each having a nominal width dimension of 66 inches				
	Total Cost	$/ft^2	$/VLF	$/floor
Cost of Stairs:	1,582,315	4.87	10,143.04	121,716.54
Cost of Photoluminous Exit Markings:	24,144	0.07	154.77	1,857.23
Total Cost:	1,606,459	4.94	10,297.81	123,573.77

Values in bold came from the "Exit Stair Cost Analysis" document.

Table 4.4 Exit Stair-Related Cost Data for Building 3: 28 Floors, Height of 336 ft (102 m), and Total Floorspace of 840 000 ft^2 (78 038 m^2)

Part 1.1: 2 exit stairs each having a nominal width dimension of 44 inches				
	Total Cost	$/ft^2	$/VLF	$/floor
Cost of Stairs:	**1,365,608**	1.63	4,064.31	48,771.71
Cost of Photoluminous Exit Markings:	**18,107**	0.02	53.89	646.68
Total Cost:	**1,383,715**	1.65	4,118.20	49,418.39

Part 1.2: 2 exit stairs each having a nominal width dimension of 56 inches				
	Total Cost	$/ft^2	$/VLF	$/floor
Cost of Stairs:	**1,548,763**	1.84	4,609.41	55,312.96
Cost of Photoluminous Exit Markings:	23,045	0.03	68.59	823.05
Total Cost:	1,571,808	1.87	4,678.00	56,136.01

Part 1.3: 2 exit stairs each having a nominal width dimension of 66 inches				
	Total Cost	$/ft^2	$/VLF	$/floor
Cost of Stairs:	1,701,392	2.03	5,063.67	60,764.01
Cost of Photoluminous Exit Markings:	27,161	0.03	80.83	970.02
Total Cost:	1,728,553	2.06	5,144.50	61,734.02

Part 2.1: 3 exit stairs each having a nominal width dimension of 44 inches				
	Total Cost	$/ft^2	$/VLF	$/floor
Cost of Stairs:	**2,048,412**	2.44	6,096.46	73,157.57
Cost of Photoluminous Exit Markings:	27,161	0.03	80.83	970.02
Total Cost:	**2,075,573**	2.47	6,177.30	74,127.59

Part 2.2: 3 exit stairs each having a nominal width dimension of 56 inches				
	Total Cost	$/ft^2	$/VLF	$/floor
Cost of Stairs:	2,323,145	2.77	6,914.12	82,969.45
Cost of Photoluminous Exit Markings:	34,568	0.04	102.88	1,234.57
Total Cost:	2,357,712	2.81	7,017.00	84,204.01

Part 2.3: 3 exit stairs each having a nominal width dimension of 66 inches				
	Total Cost	$/ft^2	$/VLF	$/floor
Cost of Stairs:	2,552,088	3.04	7,595.50	91,146.01
Cost of Photoluminous Exit Markings:	40,741	0.05	121.25	1,455.03
Total Cost:	2,592,829	3.09	7,716.75	92,601.04

Part 3.1: 4 exit stairs each having a nominal width dimension of 44 inches				
	Total Cost	$/ft^2	$/VLF	$/floor
Cost of Stairs:	**2,731,216**	3.25	8,128.62	97,543.43
Cost of Photoluminous Exit Markings:	36,214	0.04	107.78	1,293.36
Total Cost:	2,767,430	3.29	8,236.40	98,836.79

Part 3.2: 4 exit stairs each having a nominal width dimension of 56 inches				
	Total Cost	$/ft^2	$/VLF	$/floor
Cost of Stairs:	3,097,526	3.69	9,218.83	110,625.93
Cost of Photoluminous Exit Markings:	46,091	0.05	137.17	1,646.09
Total Cost:	3,143,617	3.74	9,356.00	112,272.02

Part 3.3: 4 exit stairs each having a nominal width dimension of 66 inches				
	Total Cost	$/ft^2	$/VLF	$/floor
Cost of Stairs:	3,402,784	4.05	10,127.33	121,528.01
Cost of Photoluminous Exit Markings:	54,321	0.06	161.67	1,940.04
Total Cost:	3,457,105	4.12	10,289.00	123,468.05

Values in bold came from the "Exit Stair Cost Analysis" document.

Table 4.5 Exit Stair-Related Cost Data for Building 4: 42 Floors, Height of 504 ft (154 m), and Total Floorspace of 1 680 000 ft² (156 076 m²)

Part 1.1: 2 exit stairs each having a nominal width dimension of 44 inches				
	Total Cost	$/ft²	$/VLF	$/floor
Cost of Stairs:	**2,044,813**	1.22	4,057.17	48,686.02
Cost of Photoluminous Exit Markings:	**27,496**	0.02	54.56	654.67
Total Cost:	**2,072,309**	1.23	4,111.72	49,340.69

Part 1.2: 2 exit stairs each having a nominal width dimension of 56 inches				
	Total Cost	$/ft²	$/VLF	$/floor
Cost of Stairs:	**2,320,884**	1.38	4,604.93	55,259.14
Cost of Photoluminous Exit Markings:	34,995	0.02	69.43	833.21
Total Cost:	2,355,879	1.40	4,674.36	56,092.35

Part 1.3: 2 exit stairs each having a nominal width dimension of 66 inches				
	Total Cost	$/ft²	$/VLF	$/floor
Cost of Stairs:	2,550,943	1.52	5,061.40	60,736.74
Cost of Photoluminous Exit Markings:	41,244	0.02	81.83	982.00
Total Cost:	2,592,187	1.54	5,143.23	61,718.74

Part 2.1: 3 exit stairs each having a nominal width dimension of 44 inches				
	Total Cost	$/ft²	$/VLF	$/floor
Cost of Stairs:	**3,067,220**	1.83	6,085.75	73,029.04
Cost of Photoluminous Exit Markings:	41,244	0.02	81.83	982.00
Total Cost:	**3,108,464**	1.85	6,167.59	74,011.04

Part 2.2: 3 exit stairs each having a nominal width dimension of 56 inches				
	Total Cost	$/ft²	$/VLF	$/floor
Cost of Stairs:	3,481,326	2.07	6,907.39	82,888.71
Cost of Photoluminous Exit Markings:	52,492	0.03	104.15	1,249.82
Total Cost:	3,533,818	2.10	7,011.54	84,138.53

Part 2.3: 3 exit stairs each having a nominal width dimension of 66 inches				
	Total Cost	$/ft²	$/VLF	$/floor
Cost of Stairs:	3,826,415	2.28	7,592.09	91,105.11
Cost of Photoluminous Exit Markings:	61,866	0.04	122.75	1,473.00
Total Cost:	3,888,281	2.31	7,714.84	92,578.11

Part 3.1: 4 exit stairs each having a nominal width dimension of 44 inches				
	Total Cost	$/ft²	$/VLF	$/floor
Cost of Stairs:	**4,089,626**	2.43	8,114.34	97,372.05
Cost of Photoluminous Exit Markings:	54,992	0.03	109.11	1,309.33
Total Cost:	4,144,618	2.47	8,223.45	98,681.38

Part 3.2: 4 exit stairs each having a nominal width dimension of 56 inches				
	Total Cost	$/ft²	$/VLF	$/floor
Cost of Stairs:	4,641,768	2.76	9,209.86	110,518.29
Cost of Photoluminous Exit Markings:	69,990	0.04	138.87	1,666.42
Total Cost:	4,711,758	2.80	9,348.73	112,184.71

Part 3.3: 4 exit stairs each having a nominal width dimension of 66 inches				
	Total Cost	$/ft²	$/VLF	$/floor
Cost of Stairs:	5,101,886	3.04	10,122.79	121,473.48
Cost of Photoluminous Exit Markings:	82,488	0.05	163.67	1,964.00
Total Cost:	5,184,374	3.09	10,286.46	123,437.48

Values in bold came from the "Exit Stair Cost Analysis" document.

Table 4.6 Exit Stair-Related Cost Data for Building 5: 75 Floors, Height of 900 ft (274 m), and Total Floorspace of 3 375 000 ft^2 (313 545 m^2)

Part 1.1: 2 exit stairs each having a nominal width dimension of 44 inches				
	Total Cost	$/ft^2	$/VLF	$/floor
Cost of Stairs:	**3,645,796**	1.08	4,050.88	48,610.61
Cost of Photoluminous Exit Markings:	**49,627**	0.01	55.14	661.69
Total Cost:	**3,695,423**	1.09	4,106.03	49,272.31

Part 1.2: 2 exit stairs each having a nominal width dimension of 56 inches				
	Total Cost	$/ft^2	$/VLF	$/floor
Cost of Stairs:	**4,140,885**	1.23	4,600.98	55,211.80
Cost of Photoluminous Exit Markings:	63,162	0.02	70.18	842.16
Total Cost:	4,204,047	1.25	4,671.16	56,053.96

Part 1.3: 2 exit stairs each having a nominal width dimension of 66 inches				
	Total Cost	$/ft^2	$/VLF	$/floor
Cost of Stairs:	4,553,459	1.35	5,059.40	60,712.79
Cost of Photoluminous Exit Markings:	74,441	0.02	82.71	992.54
Total Cost:	4,627,900	1.37	5,142.11	61,705.33

Part 2.1: 3 exit stairs each having a nominal width dimension of 44 inches				
	Total Cost	$/ft^2	$/VLF	$/floor
Cost of Stairs:	**5,468,694**	1.62	6,076.33	72,915.92
Cost of Photoluminous Exit Markings:	74,441	0.05	82.71	992.54
Total Cost:	**5,543,135**	1.67	6,159.04	73,908.46

Part 2.2: 3 exit stairs each having a nominal width dimension of 56 inches				
	Total Cost	$/ft^2	$/VLF	$/floor
Cost of Stairs:	6,211,328	1.84	6,901.48	82,817.70
Cost of Photoluminous Exit Markings:	94,742	0.03	105.27	1,263.23
Total Cost:	6,306,070	1.87	7,006.74	84,080.93

Part 2.3: 3 exit stairs each having a nominal width dimension of 66 inches				
	Total Cost	$/ft^2	$/VLF	$/floor
Cost of Stairs:	6,830,189	2.02	7,589.10	91,069.18
Cost of Photoluminous Exit Markings:	111,661	0.03	124.07	1,488.81
Total Cost:	6,941,850	2.06	7,713.17	92,557.99

Part 3.1: 4 exit stairs each having a nominal width dimension of 44 inches				
	Total Cost	$/ft^2	$/VLF	$/floor
Cost of Stairs:	**7,291,592**	2.16	8,101.77	97,221.23
Cost of Photoluminous Exit Markings:	99,254	0.03	110.28	1,323.39
Total Cost:	7,390,846	2.19	8,212.05	98,544.61

Part 3.2: 4 exit stairs each having a nominal width dimension of 56 inches				
	Total Cost	$/ft^2	$/VLF	$/floor
Cost of Stairs:	8,281,770	2.45	9,201.97	110,423.60
Cost of Photoluminous Exit Markings:	126,323	0.04	140.36	1,684.31
Total Cost:	8,408,093	2.49	9,342.33	112,107.91

Part 3.3: 4 exit stairs each having a nominal width dimension of 66 inches				
	Total Cost	$/ft^2	$/VLF	$/floor
Cost of Stairs:	9,106,918	2.70	10,118.80	121,425.58
Cost of Photoluminous Exit Markings:	148,881	0.04	165.42	1,985.08
Total Cost:	9,255,799	2.74	10,284.22	123,410.66

Values in bold came from the "Exit Stair Cost Analysis" document.

4.2.2 Incremental Cost of Adding an Exit Stair

The cost estimates reported in Tables 4.2 through 4.6 are easily reformatted to examine the incremental cost of adding an exit stair. The case of adding an exit stair is of special interest because for buildings greater than 420 ft (128 m) in height the recently adopted provisions of the International Building Code (IBC) allow the substitution of occupant evacuation elevators in lieu of an additional exit stair.[35] The case for examining exit stairs of nominal width of 66 in (168 cm), rather than limiting our attention to 44 in (112 cm) and 56 in (142 cm) exit stairs is of special interest because a recently adopted provision of the IBC requires an increase of 50 % in the width of exit stairs in new sprinklered buildings.[36] It is important to note that the incremental cost of adding an exit stair differs from the total cost of an exit stair. The incremental cost of adding an exit stair reflects the substitution of the exit stair "footprint" for a corresponding floorspace footprint. Thus, it is necessary to net out the cost of the floorspace footprint. This is done by subtracting a nominal rate of $100.00 per ft^2 ($1076.43 per m^2) from the total cost of the exit stair.[37]

Table 4.7 reports estimates of the incremental cost of adding an exit stair for each of the three exit stair widths analyzed. The incremental costs reported in Table 4.7 are on a per exit stair basis; they are initial capital costs referred to hereafter as first costs. Table 4.8 reports estimates of the incremental cost on a per floor basis. Reference to Tables 4.7 and 4.8 reveals that incremental costs decline as the width of the exit stair is increased for all five prototypical buildings. This is because there is a substantial fixed cost associated with an exit stair (e.g., framing, landings, doors), whereas increasing the stair width from 44 in (112 cm) to 66 in (168 cm) results in a less than proportionate increase in total cost. Thus, when the fixed nominal rate of $100.00 per ft^2 ($1076.43 per m^2) is subtracted from the total cost of the stair, the incremental cost declines as the stair width increases.

Table 4.7 Incremental Cost of Adding an Exit Stair

Width of Stairs (in)	Change in Total Cost				
	Building #1 5 Floors	Building #2 13 Floors	Building #3 28 Floors	Building #4 42 Floors	Building #5 75 Floors
44	45,728	113,669	241,058	359,955	640,212
56	32,846	82,466	175,504	262,339	467,023
66	18,862	48,015	102,676	153,694	273,950

---

[35] *International Building Code*, Sections 403.5.2 and 3008.4, *Op Cit.*
[36] National Institute of Standards and Technology. "Safer Buildings Are Goal of New Code Changes Based on Recommendations from NIST World Trade Center Investigation" TechBeat: October 1, 2008. http://www.nist.gov/public_affairs/releases/wtc_100108.html (accessed December 2008).
[37] The nominal rate of $100.00 per ft^2 for building floorspace is based on data compiled from RS Means, CostWorks.

Table 4.8 Incremental Cost per Floor of Adding an Exit Stair

Width of Stairs (in)	Change in Total Cost per Floor per Exit Stair				
	Building #1 5 Floors	Building #2 13 Floors	Building #3 28 Floors	Building #4 42 Floors	Building #5 75 Floors
44	9,146	8,744	8,609	8,570	8,536
56	6,569	6,344	6,268	6,246	6,227
66	3,772	3,693	3,667	3,659	3,653

As noted earlier, for buildings greater than 420 ft (128 m) in height, the IBC requires either an additional exit stair or occupant evacuation elevators. Reference to Table 4.7 provides an important component of this decision problem, namely the incremental capital cost of adding an exit stair. To cover the exit stair portion of the break-even analysis, we need information on any future operations and maintenance costs associated with the installation of an additional exit stair. Although the incremental costs to heat and cool the column of air in the exit stair is not expected to exceed the office and hallway spaces around it, the "footprint" taken up by the added exit stair removes some space from a revenue generating status (i.e., a loss of rental income). Figure 4.1 shows the plan and cross section of an exit stair with a nominal width of 44 in (112 cm).

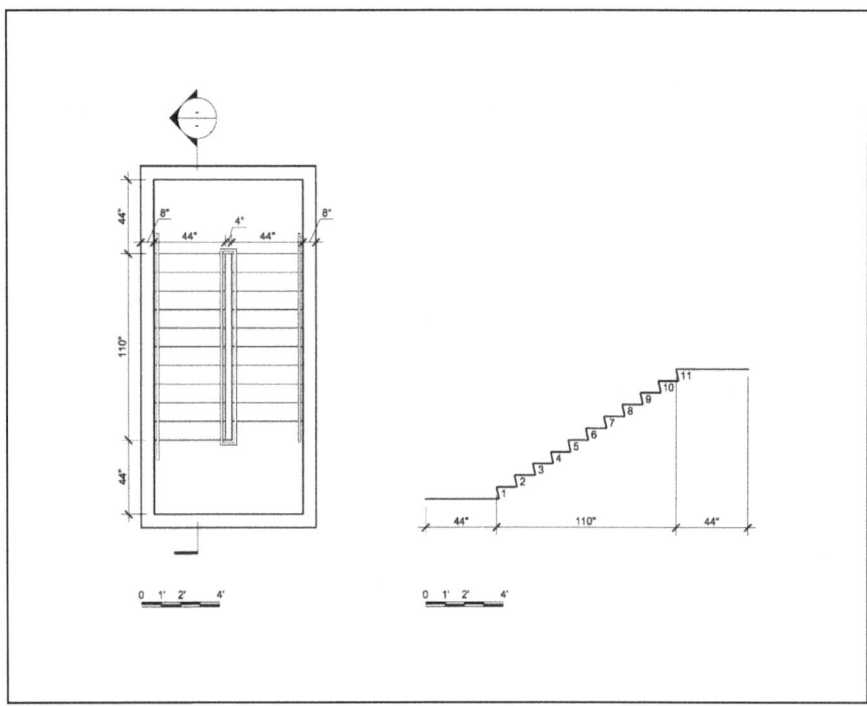

Figure 4.1 Plan and Cross Section of Exit Stair Used in Calculating Lost Rental Income: Exit Stair with Nominal Width of 44 in (112 cm)

Table 4.9 provides estimates of lost rental space. The "footprints" presented in Table 4.9 were derived from CostWorks.[38, 39] To get the value of the lost rental stream, we use information from the Building Owners and Managers Association (BOMA) Experience Exchange Report.[40] A rental rate of $36.92 per ft² ($397.37 per m²) is used to estimate the annual lost rental income. This rate corresponds to the rate in downtown Washington, DC for buildings with over 600 000 ft² (55 740 m²) of floorspace in July 2007.[41] The expected annual loss of rental income values are reported in Table 4.10. For prototypical buildings 4 and 5—the two buildings greater than 420 ft (128 m) in height—they range from $248 878 per year for Building 4 with an exit stair of nominal width 44 in (112 cm) to $754 091 for Building 5 with an exit stair of nominal width 66 in (168 cm). Rental rates for major urban areas vary from a high of $43.18 per ft² ($464.80 per m²) for downtown New York to a low of $18.18 per ft² ($195.69 per m²) for downtown Houston.[42, 43]

---

[38] RS Means, CostWorks.

[39] CostWorks contains information on typical ranges of risers for various story heights. The prototypical buildings in Table 4.1 have a story height of 12 feet. For a story height of 12 feet, the average number of risers is 21. The minimum tread width for commercial buildings is 11 inches. The cross sectional area in the calculation of loss of rental space includes two landing areas, two stair areas, a gap area between stairs, and surrounding wall space. One flight of stairs would have 11 risers. The other flight of stairs would have 10 risers. The landing area would be slightly larger on the side where there are 10 risers. The increase in the landing area is the same as the decrease in the stair area. For the calculation, we would use 11 risers to calculate the length of stairs with no increase in the landing area. For the case where the stair width is 44 inches, the stair length is 11 inches x (11-1) = 110 inches. In this calculation, 1 is subtracted from 11 because the last step of a flight of stairs is on the landing area. The gap between the stairs in our calculation is 4 inches. The wall is 8 inches in thickness. With a stair width of 44 inches, the width of a landing area is also 44 inches. Therefore, the length of the cross sectional area is 8 inches x 2 + 44 inches x 2 + 110 inches = 214 inches. The width of the cross sectional area is 8 inches x 2 + 44 inches x 2 + 4 inches = 108 inches. The cross sectional area is therefore 214 inches x 108 inches = 23 112 square inches = 160.5 square feet. In the case of prototypical building #4, where there are 42 floors, the loss of rental space due to an additional stairwell is 160.5 square feet x 42 = 6741 square feet.

[40] BOMA, Experience Exchange Report.

[41] Since the cost estimates contained in the Exit Stair Cost Analysis document are for July 2007 in Washington, DC, all other cost information presented in this report uses July 2007 as its reference date and the *greater Washington, DC metropolitan area* as the building's location.

[42] BOMA, Experience Exchange Report.

[43] Rental rates for downtown locations for the 10 most populated urban areas, listed according to population, are as follows: (1) New York, $43.18; (2) Los Angeles, $25.03; (3) Chicago, $25.69; (4) Dallas, $20.09; (5) Philadelphia, $27.02; (6) Houston, $18.18; (7) Miami, $30.02; (8) Washington, DC, $36.92; (9) Atlanta, $26.47; and (10) Boston, $35.95.

Table 4.9 Loss of Rental Space Due to the Installation of an Additional Exit Stair

Building	Height (ft)	Floors	Square Feet Per Floor	Stair Width (in)	Crosssectional Area (sq ft)	Total Loss of Rental Space (sq ft)
1	60	5	20,000	44	161	803
1	60	5	20,000	56	218	1,091
1	60	5	20,000	66	272	1,362
2	156	13	25,000	44	161	2,087
2	156	13	25,000	56	218	2,836
2	156	13	25,000	66	272	3,540
3	336	28	30,000	44	161	4,494
3	336	28	30,000	56	218	6,109
3	336	28	30,000	66	272	7,625
4	504	42	40,000	44	161	6,741
4	504	42	40,000	56	218	9,163
4	504	42	40,000	66	272	11,438
5	900	75	45,000	44	161	12,038
5	900	75	45,000	56	218	16,363
5	900	75	45,000	66	272	20,425

Table 4.10 Annual Loss of Rental Income Due to the Installation of an Additional Exit Stair

Building	Floors	Stair Width (in)	Total Loss of Rental Space (sq ft)	Annual Loss of Rental Income in Dollars	Annual Loss of Rental Income in Dollars per Floor
1	5	44	803	29,647	5,929
1	5	56	1,091	40,280	8,056
1	5	66	1,362	50,285	10,057
2	13	44	2,087	77,052	5,927
2	13	56	2,836	104,705	8,054
2	13	66	3,540	130,697	10,054
3	28	44	4,494	165,918	5,926
3	28	56	6,109	225,544	8,055
3	28	66	7,625	281,515	10,054
4	42	44	6,741	248,878	5,926
4	42	56	9,163	338,298	8,055
4	42	66	11,438	422,291	10,055
5	75	44	12,038	444,425	5,926
5	75	56	16,363	604,104	8,055
5	75	66	20,425	754,091	10,055

The life-cycle cost analysis presented in this report uses 2007 as the base year, a 25-year study period, and a 2.7 % real discount rate. The 2.7 % real discount rate conforms to OMB guidance for cost-effectiveness analyses of government programs with either a 20-year or 30-year study period.[44, 45] The first cost and annual recurring cost estimates are shown in Tables 4.7 and 4.10. The length of the study period is based on the expected service life of elevators reported by Whitestone Research.[46]

The results of the life-cycle cost analyses are reported in Table 4.11. For each building, three life-cycle cost analyses were performed—one for each exit stair width. The columns of Table 4.11 are laid out so that the values for the incremental first cost can be easily traced to Table 4.7 and the annual loss of rental income can be easily traced to Table 4.10. The last three columns of Table 4.11 contain the present value of the annual stream of lost rental income, the life-cycle cost for each building exit stair width combination, and the corresponding life-cycle cost per floor. The value of life-cycle cost found in the nest to last column is the sum of the incremental first cost and the present value of the annual stream of lost rental income.

---

[44] Office of Management and Budget. "Circular A-94: Guidelines and Discount Rates for Benefit-Cost Analysis of Federal Programs."
[45] Orszag, P.R. "Memorandum for the Heads of Departments and Agencies: 2010 Discount Rates for OMB Circular No. A-94," Office of Management and Budget, December 8, 2009.
[46] Whitestone Research, 2008. "The Whitestone Building Maintenance and Repair Cost Reference: 2008-2009." 13th Edition. Santa Barbara, CA: Whitestone Research.

Table 4.11 Life-Cycle Costs of an Additional Exit Stair

Building	Floors	Stair Width (in)	Incremental First Cost in Dollars	Loss of Rental Income		Life-Cycle Cost in Dollars	Life-Cycle Cost in Dollars Per Floor
				Annual Value in Dollars	Present Value in Dollars		
1	5	44	45,728	29,628	533,601	579,328	115,866
1	5	56	32,846	40,274	725,320	758,166	151,633
1	5	66	18,862	50,273	905,403	924,265	184,853
2	13	44	113,669	77,034	1,387,361	1,501,030	115,464
2	13	56	82,466	104,711	1,885,832	1,968,298	151,408
2	13	66	48,015	130,709	2,354,048	2,402,063	184,774
3	28	44	241,058	165,918	2,988,163	3,229,220	115,329
3	28	56	175,504	225,532	4,061,791	4,237,296	151,332
3	28	66	102,676	281,527	5,070,258	5,172,934	184,748
4	42	44	359,955	248,878	4,482,244	4,842,199	115,290
4	42	56	262,339	338,298	6,092,687	6,355,027	151,310
4	42	66	153,694	422,291	7,605,387	7,759,080	184,740
5	75	44	640,212	444,425	8,004,008	8,644,219	115,256
5	75	56	467,023	604,104	10,879,799	11,346,822	151,291
5	75	66	273,950	754,091	13,581,047	13,854,997	184,733

The next to last column of Table 4.11 contains the break-even values indicating when the installation of occupant evacuation elevators is cost effective vis-à-vis the addition of an exit stair. Recall that in a life-cycle cost analysis the alternative with the lowest life-cycle cost is the most cost effective. Thus, for a given prototypical building/stair width combination, occupant evacuation elevators are cost effective *if and only if* their corresponding life-cycle cost is less than or equal to the numbers in the next to last column of Table 4.11. For example, if Building 4 were a new sprinklered building (i.e., a building requiring an exit stair width of 66 in (168 cm)), then occupant evacuation elevators would be cost effective if their life-cycle costs over the 25-year study period were less than or equal to the break-even value of $7 759 080.

4.3   Occupant Evacuation Elevators and Fire Service Access Elevators

As noted earlier, for buildings greater than 420 ft (128 m) in height, the IBC requires either an additional exit stair or occupant evacuation elevators.[47] The exit stair calculations presented in the previous section (see Table 4.11) established the break-even value for the life-cycle costs of installing occupant evacuation elevators in all five prototypical buildings. Subsection 4.3.1 presents data on the incremental costs of when it would be cost-effective to "convert" a standard passenger elevator system to an occupant evacuation elevator system.[48] Although the focus is in on Buildings 4 and 5 when comparing the cost-effectiveness of occupant evacuation elevators vis-à-vis the installation of an additional exit stair, cost data for occupant evacuation elevator systems for Buildings 2 and 3 are also included. This is because the U.S. General Services Administration (GSA) is considering adding a provision in the GSA Design Standards[49] to require occupant evacuation elevators in buildings greater than 120 ft (37 m) in height.[50] Both Buildings 2 (13 floors) and Building 3 (28 floors) are greater than 120 ft (37 m) in height.

Much of the focus of this report is on buildings with 11 or more floors (see Section 2.2). By and large, these buildings are over 120 ft (37 m) in height. Buildings with 11 or more floors are of particular importance to any comprehensive economic analysis of egress and life-safety measures because a recently adopted provision in the IBC requires a minimum of one fire service access elevator for buildings more than 120 ft (37 m) in height.[51] Fire service access elevators are an important consideration in the analysis of egress

---

[47] Passenger elevators must meet specific criteria to be used for occupant evacuation purposes; these criteria are provided in Section 3008 of the *2009 International Building Code, Op Cit.*

[48] The term "convert" is meant to convey the idea that there are additional costs associated with occupant evacuation elevators that are over and above those needed to install a standard passenger elevator.

[49] U.S. General Services Administration, Public Building Service. 2005. *Facilities Standards for the Public Buildings Service*, PBS-100. (Washington, DC: U.S. General Services Administration, March 2005).

[50] David Frable, Senior Fire Protection Engineer, U.S. General Services Administration, Personal Communication, August 3, 2010.

[51] Elevators must meet specific criteria to be used for fire service access; these criteria are provided in Section 3007 of the *2009 International Building Code, Op Cit.*

alternatives, as responder use of stairs has been shown to increase total building evacuation time.[52] Section 4.3.2 presents these data for Buildings 2, 3, 4, and 5.

4.3.1 Occupant Evacuation Elevators

As was the case for the exit stair cost analysis, cost information presented in this subsection uses July 2007 as its reference date and the greater Washington, DC metropolitan area as the building's location. Cost data were compiled from elevator contractors, designers, cost estimating guides, and subject matter experts. The incremental costs of converting a standard passenger elevator system to an occupant evacuation elevator system are summarized in Tables 4.12 through 4.15. The tables are divided into two parts: Part A summarizes the incremental investment costs (i.e., first costs) and Part B summarizes the annual recurring costs due to the additional maintenance requirements associated with occupant evacuation elevators. Table 4.12 summarizes data for Building 2. Table 4.13 summarizes data for Building 3. Table 4.14 summarizes data for Building 4. Table 4.15 summarizes data for Building 5. It is important to note that the data contained in Tables 4.12 through 4.15 are an aggregation of a series of cost estimating relationships. Readers wishing to examine the specifications, assumptions, and cost estimating relationships for Buildings 2, 3, 4, and 5 are referred to Appendix A.

Four types of cost data are presented in Tables 4.12 through 4.15: total cost of the system as configured; the cost per unit of floor area; the cost per unit of building height; and the cost per floor. Specific cost items reported in Tables 4.12 through 4.15 are: the cost of water protection; the cost of installing signage, lobby status indicator, and a two-way communication system; the costs of protecting wiring and cables; the costs of enclosing the lobbies on all floors above the ground floor; and the total first cost of the occupant evacuation elevator system. Building 5 includes a sky lobby, which is used to provide access from the ground floor to the high-zone elevators, via shuttle elevators. Since the shuttle elevators to the sky lobby and the high-zone elevators are treated as two lobbies, a connection between these lobbies is needed and the connections need to be protected. Hence Table 4.15 includes an additional cost item: sky lobby increment. These first costs are recorded in Part A of each table. The annual recurring costs associated with elevator maintenance are recorded in Part B of each table. Part B of Table 4.15 includes an additional recurring cost item: lost rental income due to sky lobby increment.

The cost estimates reported in Tables 4.12 through 4.15 are easily reformatted to facilitate calculation of the life-cycle costs of converting a standard passenger elevator system to an occupant evacuation elevator system. As noted earlier, the life-cycle cost analysis presented in this report uses 2007 as the base year, a 25-year study period, and a 2.7 % real discount rate. The first cost and annual recurring cost estimates are shown in

---

[52] Averill, J.D., and W. Song. 2007. *Accounting for Emergency Response in Building Evacuation: Modeling Differential Egress Capacity Solutions*. NISTIR 7425. Gaithersburg, MD: National Institute of Standards and Technology.

Tables 4.12 through 4.15. The length of the study period is based on the expected service life of elevators reported by Whitestone Research.[53]

Table 4.12 Incremental Cost of Converting Passenger Elevators to Occupant Evacuation Elevators for Building 2: 13 Floors, Height of 156 ft (48 m), and Total Floorspace of 325 000 ft^2 (30 193 m^2)

Part A: Incremental Capital Cost of Converting Passenger Elevators to Occupant Evacuation Elevators for Building 2

Cost of Coverting Standard Elevators to Occupant Evacuation Elevators				
	Total Cost	$/ft^2	$/vlf	$/floor
Water Protection	97,500	0.30	625.00	7,500.00
Signage, Lobby Status Indicator, and Two-Way Communication	54,000	0.17	346.15	4,153.85
Protection of Wiring/Cables	1,800	0.01	11.54	138.46
Lobby Enclosure	300,000	0.92	1,923.08	23,076.92
Total	453,300	1.39	2,905.77	34,869.23

Part B: Annually Recurring Costs of Converting Passenger Elevators to Occupant Evacuation Elevators for Building 2

Cost of Coverting Standard Elevators to Occupant Evacuation Elevators				
	Total Cost	$/ft^2	$/vlf	$/floor
Maintenance	5,400	0.02	34.62	415.38
Total	5,400	0.02	34.62	415.38

---

[53] Whitestone Research, 2008. *Op cit.*

Table 4.13 Incremental Cost of Converting Passenger Elevators to Occupant Evacuation Elevators for Building 3: 28 Floors, Height of 336 ft (102 m), and Total Floorspace of 840 000 ft² (78 038 m²)

Part A: Incremental Capital Cost of Converting Passenger Elevators to Occupant Evacuation Elevators for Building 3

Cost of Coverting Standard Elevators to Occupant Evacuation Elevators				
	Total Cost	$/ft²	$/vlf	$/floor
Water Protection	342,000	0.41	1,017.86	12,214.29
Signage, Lobby Status Indicator, and Two-Way Communication	121,500	0.14	361.61	4,339.29
Protection of Wiring/Cables	4,800	0.01	14.29	171.43
Lobby Enclosure	675,000	0.80	2,008.93	24,107.14
Total	1,143,300	1.36	3,402.68	40,832.14

Part B: Annually Recurring Costs of Converting Passenger Elevators to Occupant Evacuation Elevators for Building 3

Cost of Coverting Standard Elevators to Occupant Evacuation Elevators				
	Total Cost	$/ft²	$/vlf	$/floor
Maintenance	14,400	0.02	42.86	514.29
Total	14,400	0.02	42.86	514.29

Table 4.14 Incremental Cost of Converting Passenger Elevators to Occupant Evacuation Elevators for Building 4: 42 Floors, Height of 504 ft (154 m), and Total Floorspace of 1 680 000 ft² (156 076 m²)

Part A: Incremental Capital Cost of Converting Passenger Elevators to Occupant Evacuation Elevators for Building 4

Cost of Coverting Standard Elevators to Occupant Evacuation Elevators				
	Total Cost	$/ft²	$/vlf	$/floor
Water Protection	702,000	0.42	1,392.86	16,714.29
Signage, Lobby Status Indicator, and Two-Way Communication	184,500	0.11	366.07	4,392.86
Protection of Wiring/Cables	9,600	0.01	19.05	228.57
Lobby Enclosure	1,025,000	0.61	2,033.73	24,404.76
Total	1,921,100	1.14	3,811.71	45,740.48

Part B: Annually Recurring Costs of Converting Passenger Elevators to Occupant Evacuation Elevators for Building 4

Cost of Coverting Standard Elevators to Occupant Evacuation Elevators				
	Total Cost	$/ft²	$/vlf	$/floor
Maintenance	28,800	0.02	57.14	685.71
Total	28,800	0.02	57.14	685.71

Table 4.15 Incremental Cost of Converting Passenger Elevators to Occupant Evacuation Elevators for Building 5: 75 Floors, Height of 900 ft (274 m), and Total Floorspace of 3 375 000 ft^2 (313 545 m^2)

Part A: Incremental Capital Cost of Converting Passenger Elevators to Occupant Evacuation Elevators for Building 5

Cost of Coverting Standard Elevators to Occupant Evacuation Elevators				
	Total Cost	$/ft^2	$/vlf	$/floor
Water Protection	1,472,000	0.44	1,635.56	19,626.67
Signage, Lobby Status Indicator, and Two-Way Communication	342,000	0.10	380.00	4,560.00
Protection of Wiring/Cables	21,900	0.01	24.33	292.00
Lobby Enclosure	1,900,000	0.56	2,111.11	25,333.33
Sky Lobby Increment	25,000	0.01	27.78	333.33
Total	3,760,900	1.11	4,178.78	50,145.33

Part B: Annually Recurring Costs of Converting Passenger Elevators to Occupant Evacuation Elevators for Building 5

Cost of Coverting Standard Elevators to Occupant Evacuation Elevators				
	Total Cost	$/ft^2	$/vlf	$/floor
Lost Rental Income Due to Sky Lobby Increment	147,680	0.04	164.09	1,969.07
Maintenance	65,700	0.02	73.00	876.00
Total	213,380	0.06	237.09	2,845.07

The results of the life-cycle cost analyses are reported in Table 4.16. The columns of Table 4.16 are laid out so that the values for the incremental first cost can be easily traced to Part A of Tables 4.12 through 4.15 and the incremental annual cost can be easily traced to Part B of Tables 4.12 through 4.15. The last three columns of Table 4.16 contain the present value of the annual stream of increased maintenance costs, the life-cycle cost for converting a standard passenger elevator system to an occupant evacuation elevator system for each prototypical building, and the corresponding life-cycle cost per floor. The value of life-cycle cost found in the next to last column is the sum of the incremental first cost and the present value of the incremental annual cost.

Table 4.16 Life-Cycle Costs of Converting a Standard Passenger Elevator System to an Occupant Evacuation Elevator System

Building	Floors	Incremental First Cost in Dollars	Incremental Annual Cost		Life-Cycle Cost in Dollars	Life-Cycle Cost in Dollars Per Floor
			Annual Value in Dollars	Present Value in Dollars		
2	13	453,300	5,400	97,253	550,553	42,350
3	28	1,143,300	14,400	259,341	1,402,641	50,094
4	42	1,921,100	28,800	518,683	2,439,783	58,090
5	75	3,760,900	213,380	3,842,937	7,603,837	101,384

Note: The annual dollar values for Building 5 include lost rental income due to the Sky Lobby Increment.

4.3.2 Fire Service Access Elevators

As was the case for the occupant evacuation elevators cost analysis, cost information presented in this subsection uses July 2007 as its reference date and the greater Washington, DC metropolitan area as the building's location. Cost data were compiled from elevator contractors, designers, cost estimating guides, and subject matter experts. The incremental costs of converting a standard service elevator to a fire service access elevator are summarized in Tables 4.17 through 4.20. The tables are divided into two parts: Part A summarizes the incremental investment costs (i.e., first costs) and Part B summarizes the annual recurring costs due to the additional maintenance requirements associated with fire service access elevators. Table 4.17 summarizes data for Building 2. Table 4.18 summarizes data for Building 3. Table 4.19 summarizes data for Building 4. Table 4.20 summarizes data for Building 5. It is important to note that the data contained in Tables 4.17 through 4.20 are an aggregation of a series of cost estimating relationships. Readers wishing to examine the specifications, assumptions, and cost estimating relationships are referred to Appendix B.

Four types of cost data are presented in Tables 4.17 through 4.20: total cost of the system as configured; the cost per unit of floor area; the cost per unit of building height; and the cost per floor. Specific cost items reported in Tables 4.17 through 4.20 are: the cost of water protection; the cost of installing signage, lobby status indicator, and a two-way communication system; the costs of protecting wiring and cables; the costs of enclosing the lobbies on all floors above the ground floor; and the total first cost of the fire service access elevator system. These first costs are recorded in Part A of each table. The annual recurring costs associated with elevator maintenance are recorded in Part B of each table.

Table 4.17 Incremental Cost of Converting Service Elevators to Fire Service Access Elevators for Building 2: 13 Floors, Height of 156 ft (48 m), and Total Floorspace of 325 000 ft² (30 193 m²)

Part A: Incremental Capital Costs of Converting Service Elevators to Fire Service Access Elevators for Building 2

Costs of Converting Service Elevators to Fire Service Access Elevators				
	Total Cost	$/ft²	$/vlf	$/floor
Water Protection	32,500	0.10	208.33	2,500.00
Signage, Lobby Status Indicator, and Two-Way Communication	54,000	0.17	346.15	4,153.85
Protection of Wiring/Cables	600	0.00	3.85	46.15
Lobby Enclosure	75,000	0.23	480.77	5,769.23
Total	162,100	0.50	1,039.10	12,469.23

Part B: Annual Recurring Costs of Converting Service Elevators to Fire Service Access Elevators for Building 2

Costs of Converting Service Elevators to Fire Service Access Elevators				
	Total Cost	$/ft²	$/vlf	$/floor
Maintenance	1,800	0.01	11.54	138.46
Total	1,800	0.01	11.54	138.46

Table 4.18 Incremental Cost of Converting Service Elevators to Fire Service Access Elevators for Building 3: 28 Floors, Height of 336 ft (102 m), and Total Floorspace of 840 000 ft² (78 038 m²)

Part A: Incremental Capital Costs of Converting Service Elevators to Fire Service Access Elevators for Building 3

Cost of Converting Service Elevators to Fire Service Access Elevators				
	Total Cost	$/ft²	$/vlf	$/floor
Water Protection	70,000	0.08	208.33	2,500.00
Signage, Lobby Status Indicator, and Two-Way Communication	121,500	0.14	361.61	4,339.29
Protection of Wiring/Cables	600	0.00	1.79	21.43
Lobby Enclosure	168,750	0.20	502.23	6,026.79
Total	360,850	0.43	1,073.96	12,887.50

Part B: Annual Recurring Costs of Converting Service Elevators to Fire Service Access Elevators for Building 3

Cost of Converting Service Elevators to Fire Service Access Elevators				
	Total Cost	$/ft²	$/vlf	$/floor
Maintenance	1,800	0.00	5.36	64.29
Total	1,800	0.00	5.36	64.29

Table 4.19 Incremental Cost of Converting Service Elevators to Fire Service Access Elevators for Building 4: 42 Floors, Height of 504 ft (154 m), and Total Floorspace of 1 680 000 ft^2 (156 076 m^2)

Part A: Incremental Capital Costs of Converting Service Elevators to Fire Service Access Elevators for Building 4

Cost of Converting Service Elevators to Fire Service Access Elevators				
	Total Cost	$/ft^2	$/vlf	$/floor
Water Protection	161,250	0.10	319.94	3,839.29
Signage, Lobby Status Indicator, and Two-Way Communication	189,000	0.11	375.00	4,500.00
Protection of Wiring/Cables	900	0.00	1.79	21.43
Lobby Enclosure	262,500	0.16	520.83	6,250.00
Total	613,650	0.37	1,217.56	14,610.71

Part B: Annual Recurring Costs of Converting Service Elevators to Fire Service Access Elevators for Building 4

Cost of Converting Service Elevators to Fire Service Access Elevators				
	Total Cost	$/ft^2	$/vlf	$/floor
Maintenance	2,700	0.00	5.36	64.29
Total	2,700	0.00	5.36	64.29

Table 4.20 Incremental Cost of Converting Service Elevators to Fire Service Access Elevators for Building 5: 75 Floors, Height of 900 ft (274 m), and Total Floorspace of 3 375 000 ft^2 (313 545 m^2)

Part A: Incremental Capital Costs of Converting Service Elevators to Fire Service Access Elevators for Building 5

Cost of Converting Service Elevators to Fire Service Access Elevators				
	Total Cost	$/ft^2	$/vlf	$/floor
Water Protection	382,500	0.11	425.00	5,100.00
Signage, Lobby Status Indicator, and Two-Way Communication	508,500	0.15	565.00	6,780.00
Protection of Wiring/Cables	1,500	0.00	1.67	20.00
Lobby Enclosure	706,250	0.21	784.72	9,416.67
Total	1,598,750	0.47	1,776.39	21,316.67

Part B: Annual Recurring Costs of Converting Service Elevators to Fire Service Access Elevators for Building 5

Cost of Converting Service Elevators to Fire Service Access Elevators				
	Total Cost	$/ft^2	$/vlf	$/floor
Maintenance	4,500	0.00	5.00	60.00
Total	4,500	0.00	5.00	60.00

The cost estimates reported in Tables 4.17 through 4.20 are easily reformatted to facilitate calculation of the life-cycle costs of converting a standard service elevator system to a fire service access elevator system. As noted earlier, the life-cycle cost analysis presented in this report uses 2007 as the base year, a 25-year study period, and a 2.7 % real discount rate. The first cost and annual recurring cost estimates are shown in Tables 4.17 through 4.20. The length of the study period is based on the expected service life of elevators reported by Whitestone Research.[54]

The results of the life-cycle cost analyses are reported in Table 4.21. The columns of Table 4.21 are laid out so that the values for the incremental first cost can be easily traced to Part A of Tables 4.17 through 4.20 and the annual maintenance cost increment can be easily traced to Part B of Tables 4.17 through 4.20. The last two columns of Table 4.21 contain the present value of the annual stream of increased maintenance costs, the life-cycle cost for converting a standard service elevator system to a fire service access elevator system for each prototypical building, and the corresponding life-cycle cost per floor. The value of life-cycle cost found in the next to last column is the sum of the incremental first cost and the present value of the annual maintenance cost increment.

Table 4.21 Life-Cycle Costs of Converting a Standard Service Elevator System to a Fire Service Access Elevator System

| Building | Floors | Incremental First Cost in Dollars | Incremental Maintenance Cost | | Life-Cycle Cost in Dollars | Life-Cycle Cost in Dollars Per Floor |
			Annual Value in Dollars	Present Value in Dollars		
2	13	162,100	1,800	32,418	194,518	14,963
3	28	360,850	1,800	32,418	393,268	14,045
4	42	613,650	2,700	48,627	662,277	15,768
5	75	1,598,750	4,500	81,044	1,679,794	22,397

4.4 Cost-Effectiveness Analysis: Additional Exit Stair or Occupant Evacuation Elevators

This section sets the ranges of values for when occupant evacuation elevators would be cost effective relative to installing an additional exit stair and when they would not be cost effective for buildings over 420 ft (128 m) in height. This decision requires a close examination of both the incremental first cost of installing an additional exit stair and the annual value of lost rental income and the incremental first cost of converting passenger elevators to occupant evacuation elevators and the incremental annual recurring cost of

---

[54] *Ibid.*

maintaining them. Three cases are examined. The first two do not require a life-cycle cost (LCC) analysis, whereas the third does.

Case 1: First Cost of Occupant Evacuation Elevators is *less than* the First Cost of the Corresponding Additional Exit Stair *and* Annual Operations and Maintenance are *less than* the Lost Rental Income Corresponding to an Additional Exit Stair.

> Occupant Evacuation Elevators *are* Cost Effective: "If a building design or system specification has both a lower initial cost and lower future costs relative to an alternative, an LCC analysis is not needed to show that the former is the economically preferable choice."[55]

Case 2: First Cost of Occupant Evacuation Elevators is *greater than* the First Cost of the Corresponding Additional Exit Stair *and* Annual Operations and Maintenance are *greater than* the Lost Rental Income Corresponding to an Additional Exit Stair.

> Occupant Evacuation Elevators *are not* Cost Effective: Installation of an additional exit stair is the most cost-effective alternative.

Case 3: First Cost of Occupant Evacuation Elevators is *greater (less) than* the First Cost of the Corresponding Additional Exit Stair *but* Annual Operations and Maintenance are *less (greater) than* the Lost Rental Income Corresponding to an Additional Exit Stair.

> In this case a life-cycle cost analysis is needed to determine which alternative is the more cost effective. For example, if the first cost for occupant evacuation elevators is greater than the addition of an exit stair of a given width, but annual operations and maintenance costs for occupant evacuation elevators are very low, then their life-cycle cost could be significantly lower.

Decision Criterion: When using life-cycle cost analysis, the alternative with the lowest life-cycle cost is the most cost effective.

Although the life-cycle costs of either an additional exit stair or occupant evacuation elevators are likely to amount to several million dollars for both Buildings 4 and 5, they represent only a small fraction (i.e., about 0.6 % to 1.2 %) of the construction cost of those buildings. Estimates provided by William Hunt, the GSA's Chief Estimator, translated into a total project cost slightly in excess of $500 million for Building 4 (42 floors, height of 504 ft (154 m), and total floorspace of 1 680 000 ft^2 (156 076 m^2)) and slightly less than $1.2 billion for Building 5 (75 floors, height of 900 ft (274 m), and total floorspace of 3 375 000 ft^2 (313 545 m^2)).[56]

---

[55] ASTM International. "Standard Practice for Measuring Life-Cycle Costs of Buildings and Building Systems," E 917, *Annual Book of ASTM Standards: 2008*, Vol. 4.11. West Conshohocken, PA: ASTM International.
[56] Personal communication, February 5, 2009.

4.4.1 Results of the Baseline Analysis

The baseline values for the incremental first costs, annual loss of rental income, present value of lost rental income, and life-cycle costs of an additional exit stair for nominal widths of 44 in (112 cm), 56 in (142 cm), and 66 in (168 cm) for Buildings 2 through 5 are recorded in Table 4.11. The baseline values for the incremental first costs, annual maintenance cost increment, present value of the annual maintenance cost increment; and life-cycle costs of converting a standard passenger elevator system to an occupant evacuation elevator system for Buildings 2 through 5 are recorded in Table 4.16. Baseline comparisons between Tables 4.11 and 4.16 reveal that the life-cycle costs of each occupant evacuation elevators alternative is less than the corresponding additional exit stair alternative. In addition, for each alternative, the incremental first cost of an additional exit stair is less than the incremental first cost of an occupant evacuation elevator system and the annual loss of rental income associated with an additional exit stair is greater than the annual maintenance cost increment for occupant evacuation elevators. Thus, Case 3 applies and a comprehensive life-cycle cost analysis is required. In performing the life-cycle cost analysis, the additional exit stair alternative is designated the base case because its first cost is less than that of the corresponding occupant evacuation elevators alternative. Furthermore, because of the way first costs and annual recurring costs/losses differ, it is appropriate to compute additional measures of economic performance. In this subsection four baseline measures of economic performance are computed and analyzed: (1) life-cycle costs; (2) present value net savings; (3) savings-to-investment ratio; and (4) adjusted internal rate of return.

Tables 4.22A and 4.22B summarize the key life-cycle cost components for each additional exit stair alternative and the corresponding occupant evacuation elevators alternative. The table includes all investment-related costs, expressed as first costs, and all non-investment costs, expressed as either the present value of lost rental income (additional exit stair) or present value of maintenance cost (occupant evacuation elevators). Note that Tables 4.22A and 4.22B include three rows for each prototypical building, one for each stair width. For each prototypical building, varying the stair width affects both the incremental first cost and the present value of lost rental income for the additional exit stair alternative, but leaves the incremental first cost and the present value of maintenance cost unchanged for the corresponding occupant evacuation elevators alternative. The columns of Tables 4.22A and 4.22B are numbered to facilitate comparisons between the two sets of alternatives. Columns 1, 2, and 3 provide information on the building number, the number of floors, and the stair width for the additional exit stair alternative. Reference to Tables 4.22A and 4.22B show that the life-cycle cost of each additional exit stair alternative recorded in Column 6 equals the sum of the entries in Column 4 (incremental first cost) and Column 5 (present value of lost rental income). The life-cycle cost of the corresponding occupant evacuation elevators alternative recorded in Column 9 equals the sum of the entries in Column 7 (incremental first cost) and Column 8 (present value of maintenance cost). Comparison of Columns 6 and 9 demonstrates that the occupant evacuation elevator alternative is the most cost-effective choice. For example, the additional exit stair alternative for Building 4 with an exit stair width of 44 in (112 cm) has a baseline value of life-cycle cost of $4.8 million versus $2.4 million for the corresponding occupant evacuation elevators alternative (see Table 4.22A).

Additional information can be gleaned from Tables 4.22A and 4.22B, and used to calculate other baseline measures of economic performance. The information needed to calculate the present value of net savings (PVNS), savings-to-investment ratio (SIR), and adjusted internal rate of return (AIRR) associated with the corresponding sets of occupant evacuation elevators alternatives compiled from Tables 4.22A and 4.22B is summarized in Tables 4.23A, 4.23B, 4.24A, 4.24B, 4.25A and 4.25B. The first three columns in each table are copied from Tables 4.22A and 4.22B, respectively.

Table 4.22A Summary of Key Life-Cycle Cost Measures for an Additional Exit Stair and Occupant Evacuation Elevators

Building	Floors	Additional Exit Stair				Occupant Evacuation Elevators		
		Stair Width (in)	Incremental First Cost	Present Value Lost Rental Income	Life-Cycle Cost in Dollars	Incremental First Cost	Present Value Maintenance Cost	Life-Cycle Cost in Dollars
Col. (1)	Col. (2)	Col. (3)	Col. (4)	Col. (5)	Col. (6) (4) + (5)	Col. (7)	Col. (8)	Col. (9) (7) + (8)
2	13	44	113,669	1,387,361	1,501,030	453,300	97,253	550,553
2	13	56	82,466	1,885,832	1,968,298	453,300	97,253	550,553
2	13	66	48,015	2,354,048	2,402,063	453,300	97,253	550,553
3	28	44	241,058	2,988,163	3,229,220	1,143,300	259,341	1,402,641
3	28	56	175,504	4,061,791	4,237,296	1,143,300	259,341	1,402,641
3	28	66	102,676	5,070,258	5,172,934	1,143,300	259,341	1,402,641

Table 4.22B Summary of Key Life-Cycle Cost Measures for an Additional Exit Stair and Occupant Evacuation Elevators

Building	Floors	Additional Exit Stair					Occupant Evacuation Elevators		
		Stair Width (in)	Incremental First Cost	Present Value Lost Rental Income	Life-Cycle Cost in Dollars	Incremental First Cost	Present Value Maintenance Cost	Life-Cycle Cost in Dollars
Col. (1)	Col. (2)	Col. (3)	Col. (4)	Col. (5)	Col. (6) (4) + (5)	Col. (7)	Col. (8)	Col. (9) (7) + (8)
4	42	44	359,955	4,482,244	4,842,199	1,921,100	518,683	2,439,783
4	42	56	262,339	6,092,687	6,355,027	1,921,100	518,683	2,439,783
4	42	66	153,694	7,605,387	7,759,080	1,921,100	518,683	2,439,783
5	75	44	640,212	8,004,008	8,644,219	3,760,900	3,842,937	7,603,837
5	75	56	467,023	10,879,799	11,346,822	3,760,900	3,842,937	7,603,837
5	75	66	273,950	13,581,047	13,854,997	3,760,900	3,842,937	7,603,837

Note: The present value entries in Column 8 for Building 5 include lost rental income due the Sky Lobby Increment.

In Tables 4.23A and 4.23B, the life-cycle cost of each additional exit stair alternative recorded in Column 4 is copied from Column 6 of Tables 4.22A and 4.22B, and the life-cycle cost of the corresponding occupant evacuation elevators alternative recorded in Column 5 is copied from Column 9 of Tables 4.22A and 4.22B. ASTM Standard Practice E 1074 defines PVNS as the difference between the life-cycle cost of the base case and the corresponding alternative. Column 6 of Tables 4.23A and 4.23B records the PVNS associated with the corresponding sets of occupant evacuation elevators alternatives; it equals the difference between Column 4 (life-cycle cost of the base case (additional exit stair)) and Column 5 (life-cycle cost of the corresponding occupant evacuation elevators alternative). For each baseline comparison, PVNS is positive indicating that the corresponding occupant evacuation elevators alternative is cost effective.

Table 4.23A Calculation of Present Value Net Savings of Occupant Evacuation Elevators vis-à-vis an Additional Exit Stair

Building	Floors	Additional Exit Stair		Occupant Evacuation Elevators	
		Stair Width (in)	Life-Cycle Cost in Dollars	Life-Cycle Cost in Dollars	Present Value Net Savings in Dollars
Col. (1)	Col. (2)	Col. (3)	Col. (4)	Col. (5)	Col. (6) (4) - (5)
2	13	44	1,501,030	550,553	950,477
2	13	56	1,968,298	550,553	1,417,745
2	13	66	2,402,063	550,553	1,851,510
3	28	44	3,229,220	1,402,641	1,826,579
3	28	56	4,237,296	1,402,641	2,834,654
3	28	66	5,172,934	1,402,641	3,770,293

Table 4.23B Calculation of Present Value Net Savings of Occupant Evacuation Elevators vis-à-vis an Additional Exit Stair

Building	Floors	Additional Exit Stair		Occupant Evacuation Elevators	
		Stair Width (in)	Life-Cycle Cost in Dollars	Life-Cycle Cost in Dollars	Present Value Net Savings in Dollars
Col. (1)	Col. (2)	Col. (3)	Col. (4)	Col. (5)	Col. (6) (4) - (5)
4	42	44	4,842,199	2,439,783	2,402,416
4	42	56	6,355,027	2,439,783	3,915,244
4	42	66	7,759,080	2,439,783	5,319,297
5	75	44	8,644,219	7,603,837	1,040,383
5	75	56	11,346,822	7,603,837	3,742,985
5	75	66	13,854,997	7,603,837	6,251,161

Table 4.24A Calculation of Savings-to-Investment Ratio of Occupant Evacuation Elevators vis-à-vis an Additional Exit Stair

			Additional Exit Stair		Occupant Evacuation Elevators				
Building	Floors	Stair Width (in)	Incremental First Cost	Present Value Lost Rental Income	Incremental First Cost	Delta Investment Cost in Dollars	Present Value Maintenance Cost	Present Value Savings in Dollars	Savings-to-Investment Ratio
Col. (1)	Col. (2)	Col. (3)	Col. (4)	Col. (5)	Col. (6)	Col. (7) (6) - (4)	Col. (8)	Col. (9) (5) - (8)	Col. (10) (9) / (7)
2	13	44	113,669	1,387,361	453,300	339,632	97,253	1,290,108	3.80
2	13	56	82,466	1,885,832	453,300	370,834	97,253	1,788,579	4.82
2	13	66	48,015	2,354,048	453,300	405,285	97,253	2,256,795	5.57
3	28	44	241,058	2,988,163	1,143,300	902,243	259,341	2,728,821	3.02
3	28	56	175,504	4,061,791	1,143,300	967,796	259,341	3,802,450	3.93
3	28	66	102,676	5,070,258	1,143,300	1,040,624	259,341	4,810,916	4.62

Table 4.24B Calculation of Savings-to-Investment Ratio of Occupant Evacuation Elevators vis-à-vis an Additional Exit Stair

Building	Floors	Additional Exit Stair		Occupant Evacuation Elevators				
		Stair Width (in)	Incremental First Cost	Present Value Lost Rental Income	Incremental First Cost	Delta Investment Cost in Dollars	Present Value Maintenance Cost	Present Value Savings in Dollars	Savings-to-Investment Ratio
Col. (1)	Col. (2)	Col. (3)	Col. (4)	Col. (5)	Col. (6)	Col. (7) (6) - (4)	Col. (8)	Col. (9) (5) - (8)	Col. (10) (9) / (7)
4	42	44	359,955	4,482,244	1,921,100	1,561,146	518,683	3,963,561	2.54
4	42	56	262,339	6,092,687	1,921,100	1,658,761	518,683	5,574,004	3.36
4	42	66	153,694	7,605,387	1,921,100	1,767,406	518,683	7,086,704	4.01
5	75	44	640,212	8,004,008	3,760,900	3,120,689	3,842,937	4,161,071	1.33
5	75	56	467,023	10,879,799	3,760,900	3,293,877	3,842,937	7,036,862	2.14
5	75	66	273,950	13,581,047	3,760,900	3,486,950	3,842,937	9,738,111	2.79

Note: The present value entries in Column 8 for Building 5 include lost rental income due the Sky Lobby Increment.

Tables 4.24A and 4.24B contain the information needed to calculate the SIR for each corresponding occupant evacuation elevators alternative vis-à-vis the base case (additional exit stair). Columns 3, 4, and 5 are associated with each additional exit stair alternative. The entries in Columns 3 (stair width), 4 (incremental first cost), and 5 (present value of lost rental income) are copied from Columns 3, 4, and 5 of Tables 4.22A and 4.22B. Columns 6 through 10 are associated with the corresponding occupant evacuation elevators alternative. The entries in Column 6 (incremental investment cost) and Column 8 (present value maintenance cost) are copied from Column 7 and Column 8 of Tables 4.22A and 4.22B. The entries in Columns 7, 9, and 10 are calculated from the entries in the other columns. ASTM Standard Practice E 964 defines SIR of the occupant evacuation elevators alternative vis-à-vis the base case (additional exit stair) as the present value of non-investment savings divided by the present value of incremental "delta" investment costs. Reference to Tables 4.24A and 4.24B show how this calculation is performed. The present value of non-investment savings equals the difference between the present value of lost rental income for the base case (Column 5) and the present value of maintenance cost (Column 8); these entries are recorded in Column 9. The present value of the delta investment cost equals the difference between the incremental first costs of each corresponding occupant evacuation elevators alternative and the incremental first costs of each exit stair base case; these entries are recorded in Column 7. The savings-to-investment ratio for each occupant evacuation elevators alternative is the quotient of the numbers in Columns 9 and 7; these entries are recorded in Column 10 of Tables 4.24A and 4.24B. For each baseline comparison, the SIR is greater than 1.0 indicating that the corresponding occupant evacuation elevators alternative is cost effective.

Table 4.25A Calculation of Adjusted Internal Rate of Return of Occupant Evacuation Elevators vis-à-vis an Additional Exit Stair

Building	Floors	Stair Width (in)	Adjusted Internal Rate of Return (Percent)
2	13	44	8.33
2	13	56	9.37
2	13	66	10.00
3	28	44	7.35
3	28	56	8.48
3	28	66	9.19

Table 4.25B Calculation of Adjusted Internal Rate of Return of Occupant Evacuation Elevators vis-à-vis an Additional Exit Stair

Building	Floors	Stair Width (in)	Adjusted Internal Rate of Return (Percent)
4	42	44	6.60
4	42	56	7.80
4	42	66	8.57
5	75	44	3.89
5	75	56	5.87
5	75	66	7.01

Tables 4.25A and 4.25B report the AIRR for each corresponding occupant evacuation elevators alternative vis-à-vis the base case (additional exit stair). The entries in the first three columns are copied form Columns 1, 2, and 3 of Tables 4.22A and 4.22B. The entries in the last column are the AIRR expressed as a percent. Several procedures exist for calculating the AIRR, denoted as r* if expressed as a decimal and R* (i.e., r* times 100) if expressed as a percent. These procedures are derived and described in detail in the report by Chapman and Fuller.[57] The most convenient procedure for calculating the AIRR is based on its relationship to the SIR. This procedure results in a closed-form solution for r*; it is expressed mathematically as:

Where

        d   =   the discount rate expressed as a decimal; and

        L   =   the length of the study period in years.

For each baseline comparison, the AIRR over the 25-year study period is greater than the minimum attractive rate of return, which is set equal to the 2.7 % real discount rate, indicating that the corresponding occupant evacuation elevators alternative is cost effective.

---

[57] Chapman, R.E., and Fuller, S.K. 1996. *Benefits and Costs of Research: Two Case Studies in Building Technology*. NISTIR 5840. Gaithersburg, MD: National Institute of Standards and Technology.

4.4.2 Results of the Sensitivity Analysis

Because the values of many variables that enter into the baseline analysis are not known with certainty, it is advisable to select a small set of variables whose impact is likely to be substantial and subject them to a sensitivity analysis. Variations in the values of these input variables translate into the value of each outcome (e.g., the SIR) in such a manner that the impacts of uncertainty can be measured quantitatively.

Sensitivity analysis may be divided into two distinct cases: (1) deterministic; and (2) probabilistic. Deterministic sensitivity analyses are the most straightforward. Their advantage is that they are easy to apply and the results are easy to explain and understand. Their disadvantage is that they do not produce results that can be tied to probabilistic levels of significance (i.e., the probability that the SIR is less than 1.0).

For example, a deterministic sensitivity analysis might use as inputs a pessimistic value, a value based on a measure of central tendency (e.g., mean or median), and an optimistic value for the variable of interest. Then an analysis could be performed to see how each outcome (e.g., the SIR) changes as each of the three chosen values for the selected input is considered in turn, while all other input variables are maintained at their baseline values. A deterministic sensitivity analysis can also be performed on different combinations of input variables. That is, several variables are altered at once and then an outcome measure is computed.

In a probabilistic sensitivity analysis, a small set of key input variables is varied either singly or in combination according to an experimental design. In most cases, probabilistic sensitivity analyses are based on Monte Carlo or Latin Hypercube techniques. The major advantage of probabilistic sensitivity analysis is that it permits the effects of uncertainty to be rigorously analyzed. For example, not only the expected value of each economic measure can be computed but also the variability of that value. In addition, probabilistic levels of significance can be attached to the computed values of each economic measure. The disadvantages of a probabilistic sensitivity analysis are: (1) that it requires many calculations carried out according to an experimental design, and is therefore practical only when used with a computer and (2) that data on the probability distribution are generally not well known.

The approach selected for this study makes use of works by McKay, Conover, and Beckman[58] and by Harris;[59] it is based on the method of model sampling. Model sampling provides the basis for many probabilistic sensitivity analyses. Model sampling is a procedure for sampling from a stochastic process to determine, through multiple trials, the characteristics of a probability distribution.

The method of model sampling was implemented through application of the *Crystal Ball* software product.[60] This software product is an add-in for spreadsheets. The *Crystal Ball*

---

[58] McKay, M. C., W. H. Conover, and R.J. Beckman. 1979. "A Comparison of Three Methods for Selecting Values of Input Variables in the Analysis of Output from a Computer Code." *Technometrics* (Vol. 21): pp. 239-245.
[59] Harris, Carl M. 1984. *Issues in Sensitivity and Statistical Analysis of Large-Scale, Computer-Based Models*. NBS GCR 84-466. Gaithersburg, MD: National Bureau of Standards.
[60] Crystal Ball. 2007. *Crystal Ball 7.3 User Manual*. Denver, CO: Decisioneering, Inc.

software product allows the user to specify a unique probability distribution for each uncertainty variable. Specification of the experimental design involves defining which variables are to be simulated and the number of simulations. Throughout this sensitivity analysis, 10 000 simulations were run for each combination of input variables under analysis. When the *Crystal Ball* software product is executed, it randomly samples from the parent probability distribution for each input variable of interest (i.e., the input variable(s) specified by the experimental design). In this analysis, a Monte Carlo sampling approach is used.

In reality, the exact nature of the parent probability distribution for each input variable is unknown. Estimates of the parameters (e.g., mean and variance) of the parent probability distribution can be made and uncertainty can be reduced by investigation and research. However, uncertainty can never be eliminated completely. Therefore, to implement the procedure without undue attention to the characterization of the parent probability distribution, it was decided to focus on only the triangular probability distribution. The triangular distribution is widely used in simulation modeling; its specification requires three data points, the minimum value, the most-likely value, and the maximum value. The triangular distribution is used whenever the range of input values is continuous and a clustering about some central value is expected.

4.4.2.1 Uncertainty Parameters

Five inputs into the baseline analysis are identified as containing uncertainty in their estimates. Table 4.26 presents the five inputs and their associated uncertainty values. The inputs include the discount rate, the rental rate, the cost of the fire door and frame system for lobbies above the ground floor, the Sky Lobby lost rental space for Building 5, and the cost associated with signage, lobby status indicator, and a two-way communication system. Each input is designed to follow a triangular distribution. The most-likely values were used in the baseline analysis.

Table 4.26 Assumptions for the Monte Carlo Simulations

Variable	Probability Distribution	Setting & Value		
		Min	Most-Likely	Max
Discount Rate	Triangular	1.0	2.7	10.0
Rental Rate	Triangular	$ 18.18	$ 36.92	$ 43.18
Fire Door and Frame System	Triangular	$ 4,000	$ 5,250	$ 6,000
Sky Lobby Lost Rental Space	Triangular	-	4,000	8,403
Signage, Lobby Status Indicator, and Two-Way Communication System	Triangular	$ 4,000	$ 4,500	$ 5,000

The discount rate is set to follow a minimum value of 1.0 %, a most-likely value of 2.7 %, and a maximum value of 10.0 %. All discount rates are real. The most-likely value conforms to OMB's guidance of using a 2.7 % real rate for cost-effectiveness analysis of government programs, which produce private benefits over either a 20- or 30-year period.[61, 62] The minimum value approximates returns on short-term Treasury notes and bonds. The maximum value follows OMB's guidance on sensitivity analysis. The rental rate is set to follow a minimum value of $18.18 per ft^2 ($195.69 per m^2), a most-likely value of $36.92 per ft^2 ($397.37 per m^2), and a maximum value of $43.18 per ft^2 ($464.80 per m^2). The range of the rental values correspond with the minimum (Houston) and maximum (New York) rental rates for downtown locations of the 10 most populated urban areas (see footnote 40). The most-likely value is set to Washington, DC. The costs of the fire door and frame system is set to follow a minimum value of $4000, a most-likely value of $5250, and a maximum value of $6000. The amount of rentable space lost to the Sky Lobby range from a minimum of 0 ft^2 (0 m^2) to a maximum of 8403 ft^2 (781 m^2), with a most likely value of 4000 ft^2 (372 m^2). The costs associated with signage, lobby status indicator, and a two-way communication system is set to a follow a minimum value of $4000, a most-likely value of $4500, and a maximum value of $5000. The minimum and maximum values for both cost variables were selected based on inputs from industry experts. The discount rate and the rental rate affect the life-cycle costs of additional exit stairs for all buildings and occupant evacuation elevators for Building 5. The cost associated with the fire door and frame system and costs associated with signage, lobby status indicator, and two-way communication system affects the life-cycle cost of the occupant evacuation elevators. The Sky Lobby lost rental space affects the life-cycle cost of the occupant evacuation elevators in Building 5.

4.4.2.2 Monte Carlo Simulation of the Life-Cycle Costs of an Additional Exit Stair and Occupant Evacuation Elevator

Monte Carlo simulation is used to assess the sensitivity of the life-cycle costs of an additional exit stair and the occupant evacuation elevator system to changes in the five uncertainty parameters (described above). Tables 4.27A presents the mean, median, minimum, maximum, and standard deviation of the life-cycle costs for Buildings 2 and 3. Tables 4.27B presents the mean, median, minimum, maximum, and standard deviation of the life-cycle costs for Buildings 4 and 5.

For Building 2 the life-cycle costs of the exit stair alternatives range from a minimum of $0.5 million (44 in (112 cm) width) to a maximum of $3.2 million (66 in (168 cm) width). The mean life-cycle costs of the exit stair alternatives are $1.2 million, $1.5 million, and $1.8 million for the 44 in (112 cm), 56 in (142 cm), and 66 in (168 cm) widths, respectively. In contrast, the life-cycle costs of the occupant evacuation elevator range from $0.4 million to $0.6 million, with a mean cost of $0.5 million. On average, the occupant evacuation elevator is less expensive to install and operate over a 25-year period when compared to the exit stair alternatives.

For Building 3 the life-cycle costs of the exit stair alternatives range from a minimum of $1.0 million (44 in (112 cm) width) to a maximum of $6.8 million (66 in (168 cm) width). The mean life-cycle costs of the exit stair alternatives are $2.5 million, $3.2 million, and $3.9 million for

---

[61] Office of Management and Budget. "Circular A-94: Guidelines and Discount Rates for Benefit-Cost Analysis of Federal Programs."
[62] Orszag, P.R. "Memorandum for the Heads of Departments and Agencies: 2010 Discount Rates for OMB Circular No. A-94," Office of Management and Budget, December 8, 2009.

the 44 in (112 cm), 56 in (142 cm), and 66 in (168 cm) widths, respectively. In contrast, the life-cycle costs of the occupant evacuation elevator range from $1.2 million to $1.5 million, with a mean cost of $1.3 million. On average, the occupant evacuation elevator is less expensive to install and operate over a 25-year period when compared to the exit stair alternatives.

For Building 4 the life-cycle costs of the exit stair alternatives range from a minimum of $1.6 million (44 in (112 cm) width) to a maximum of $10.2 million (66 in (168 cm) width). The mean life-cycle costs of the exit stair alternatives are $3.7 million, $4.8 million, and $5.8 million for the 44 in (112 cm), 56 in (142 cm), and 66 in (168 cm) widths, respectively. In contrast, the life-cycle costs of the occupant evacuation elevator range from $2.0 million to $2.6 million, with a mean cost of $2.3 million. On average, the occupant evacuation elevator is less expensive to install and operate over a 25-year period when compared to the exit stair alternatives.

For Building 5 the life-cycle costs of the exit stair alternatives range from a minimum of $2.8 million (44 in (112 cm) width) to a maximum of $18.2 million (66 in (168 cm) width). The mean life-cycle costs are $6.6 million, $8.6 million, and $10.4 million for the 44 in (112 cm), 56 in (142 cm), and 66 in (168 cm) widths, respectively. In contrast, the life-cycle costs of the occupant evacuation elevator range from $4.3 million to $11.4 million, with a mean cost of $6.8 million. On average, the occupant evacuation elevator is less expensive to install and operate over a 25-year period when compared to the exit stair alternatives of only the 56 in (142 cm) and 66 in (168 cm) widths.

A notable difference across the life-cycle costs is costs associated with the additional exit stairs vary considerably more than the corresponding occupant evacuation elevators—i.e., there is greater uncertainty in the estimates of the life-cycle cost for stairs than elevators. This can be seen in comparisons of the standard deviation, where for instance, the standard deviation of the life-cycle cost of the 44 in (112 cm) stair in Building 4 is almost eight times larger than the corresponding life-cycle costs of the occupant evacuation elevator system.

4.4.2.3 Monte Carlo Simulation of the Economic Performance of Occupant Evacuation Elevators Compared to an Additional Exit Stair: PVNS, SIR, and AIRR

The economic performance of the occupant evacuation elevators are compared with the installation of an additional exit stair, while again, accounting for uncertainty of five key model inputs, using three economic measures: present value net savings (PVNS), savings-to-investment ratio (SIR), and adjusted internal rate of return (AIRR). Summary statistics of the economic performance measures are presented in Tables 4.28A (Buildings 2 and 3) and 4.28B (Buildings 4 and 5).

Table 4.27A Summary Statistics from the Monte Carlo Simulations of Life-Cycle Cost for an Additional Exit Stair and Occupant Evacuation Elevators (statistics in thousands of dollars)

Building	Floors	Stair Width	Additional Exit Stair					Occupant Evacuation Elevator				
			Mean	Min	Median	Max	Std. Dev.	Mean	Min	Median	Max	Std. Dev.
2	13	44	1,150	495	1,134	1,949	260					
2	13	56	1,490	601	1,469	2,578	353	527	447	527	601	26
2	13	66	1,806	695	1,779	3,163	440					
3	28	44	2,472	1,063	2,438	4,195	559					
3	28	56	3,208	1,293	3,162	5,550	760	1,343	1,152	1,344	1,522	62
3	28	66	3,888	1,497	3,830	6,812	949					

Table 4.27B Summary Statistics from the Monte Carlo Simulations of Life-Cycle Cost for an Additional Exit Stair and Occupant Evacuation Elevators (statistics in thousands of dollars)

Building	Floors	Stair Width	Additional Exit Stair					Occupant Evacuation Elevator				
			Mean	Min	Median	Max	Std. Dev.	Mean	Min	Median	Max	Std. Dev.
4	42	44	3,707	1,593	3,655	6,291	839					
4	42	56	4,811	1,938	4,742	8,324	1,140	2,330	2,004	2,330	2,644	109
4	42	66	5,832	2,245	5,745	10,218	1,423					
5	75	44	6,616	2,841	6,525	11,232	1,498					
5	75	56	8,590	3,459	8,466	14,864	2,036	6,768	4,271	6,641	11,408	1,114
5	75	66	10,414	4,009	10,258	18,245	2,542					

For Building 2, the PVNS range from a minimum of -$0.04 million (compared to the 44 in (112 cm) exit stair) to a maximum of $2.6 million (compared to the 66 in (168 cm) exit stair). The mean PVNS are all positive, at $0.6 million for the 44 in (112 cm) exit stair, $1.0 million for the 56 in (142 cm) exit stair, and $1.3 million for the 66 in (168 cm) exit stair. On average, the occupant evacuation elevators are cost-effective as compared to the additional exit stairs. Again, the cost-effectiveness is further highlighted with mean SIR's greater than one and AIRR's greater than 0.027.

While the occupant evacuation elevators are generally cost-effective, the sensitivity analysis reveals combinations of the uncertainty parameters yield comparisons that are not (e.g., a negative PVNS). For the comparison with the 44 in (112 cm) exit stair this occurred in 0.02 % of the simulations (i.e., 2 trials out of 10 000) (analysis not shown). The majority of variation in the PVNS measure (and SIR and AIRR, for that matter) can be explained by variation in the rental rate and discount rate. For the 44 in (112 cm) exit stair, 55 % of the variation in PVNS can be explained by changes in the discount rate, followed by 45 % of the variation explained by variation in rental rate (analysis not shown). Minimum PVNS correspond with low rental rates and high discount rates.

For the comparison with the 56 in (142 cm) exit stair the occupant evacuation elevators were cost-effective in all of the simulations. The majority of variation in the PVNS measure can be explained by variation in the rental rate and discount rate. For the 56 in (142 cm) exit stair, 56 % of the variation in PVNS can be explained by changes in the discount rate, followed by 44 % of the variation explained by variation in rental rate (analysis not shown). Minimum PVNS correspond with low rental rates and high discount rates.

For the comparison with the 66 in (168 cm) exit stair the occupant evacuation elevators were cost-effective in all of the simulations. The majority of variation in the PVNS measure (and SIR and AIRR, for that matter) can be explained by variation in the rental rate and discount rate. For the 66 in (168 cm) exit stair, 57 % of the variation in PVNS can be explained by changes in the discount rate, followed by 43 % of the variation explained by variation in rental rate (analysis not shown). Minimum PVNS correspond with low rental rates and high discount rates.

For Building 3, the PVNS range from a minimum of -$0.3 million (compared to the 44 in (112 cm) exit stair) to a maximum of $5.4 million (compared to the 66 in (168 cm) exit stair). The mean PVNS are all positive, at $1.1 million for the 44 in (112 cm) exit stair, $1.9 million for the 56 in (142 cm) exit stair, and $2.5 million for the 66 in (168 cm) exit stair. On average, the occupant evacuation elevators are cost-effective as compared to the additional exit stairs. Again, the cost-effectiveness is further highlighted with mean SIR's greater than one and AIRR's greater than 0.027.

While the occupant evacuation elevators are generally cost-effective, the sensitivity analysis reveals combinations of the uncertainty parameters yield comparisons that are not (e.g., a negative PVNS). For the comparison with the 44 in (112 cm) exit stair this occurred in 0.5 % of the simulations (i.e., 50 trials out of 10 000) (analysis not shown). The majority of variation in the PVNS measure (and SIR and AIRR, for that matter) can be explained by variation in the rental rate and discount rate. For the 44 in (112 cm) exit stair, 54 % of the variation in PVNS can be explained by changes in the discount rate, followed by 46 % of the variation explained by variation in rental rate (analysis not shown). Minimum PVNS correspond with low rental rates and high discount rates.

For the comparison with the 56 in (142 cm) exit stair the occupant evacuation elevators were not cost-effective in 0.02 % of the simulations (i.e., 2 trials out of 10 000) (analysis not shown). The majority of variation in the PVNS measure can be explained by variation in the rental rate and discount rate. For the 56 in (142 cm) exit stair, 55 % of the variation in PVNS can be explained by changes in the discount rate, followed by 45 % of the variation explained by variation in rental rate (analysis not shown). Minimum PVNS correspond with low rental rates and high discount rates.

For the comparison with the 66 in (168 cm) exit stair the occupant evacuation elevators were cost-effective in all of the simulations. The majority of variation in the PVNS measure (and SIR and AIRR, for that matter) can be explained by variation in the rental rate and discount rate. For the 66 in (168 cm) exit stair, 56 % of the variation in PVNS can be explained by changes in the discount rate, followed by 44 % of the variation explained by variation in rental rate (analysis not shown). Minimum PVNS correspond with low rental rates and high discount rates.

For Building 4, the PVNS range from a minimum of -$0.7 million (compared to the 44 in (112 cm) exit stair) to a maximum of $7.7 million (compared to the 66 in (168 cm) exit stair). The mean PVNS are all positive, at $1.4 million for the 44 in (112 cm) exit stair, $2.5 million for the 56 in (142 cm) exit stair, and $3.5 million for the 66 in (168 cm) exit stair. On average, the occupant evacuation elevators are cost-effective as compared to the additional exit stairs. Again, the cost-effectiveness is further highlighted with mean SIR's greater than one and AIRR's greater than 0.027.

While the occupant evacuation elevators are generally cost-effective, the sensitivity analysis reveals combinations of the uncertainty parameters yield comparisons that are not (e.g., a negative PVNS). For the comparison with the 44 in (112 cm) exit stair this occurred in 2.6 % of the simulations (i.e., 264 trials out of 10 000) (analysis not shown). The majority of variation in the PVNS measure (and SIR and AIRR, for that matter) can be explained by variation in the rental rate and discount rate. For the 44 in (112 cm) exit stair, 52 % of the variation in PVNS can be explained by changes in the discount rate, followed by 48 % of the variation explained by variation in rental rate (analysis not shown). Minimum PVNS correspond with low rental rates and high discount rates.

For the comparison with the 56 in (142 cm) exit stair the occupant evacuation elevators were not cost-effective in 0.1 % of the simulations (i.e., 13 trials out of 10 000) (analysis not shown). The majority of variation in the PVNS measure can be explained by variation in the rental rate and discount rate. For the 56 in (142 cm) exit stair, 54 % of the variation in PVNS can be explained by changes in the discount rate, followed by 46 % of the variation explained by variation in rental rate (analysis not shown). Minimum PVNS correspond with low rental rates and high discount rates.

For the comparison with the 66 in (168 cm) exit stair the occupant evacuation elevators were not cost-effective in 0.01 % of the simulations (i.e., 1 trials out of 10 000) (analysis not shown). The majority of variation in the PVNS measure (and SIR and AIRR, for that matter) can be explained by variation in the rental rate and discount rate. For the 66 in (168 cm) exit stair, 55 % of the variation in PVNS can be explained by changes in the discount rate, followed by 45 % of the variation explained by variation in rental rate (analysis not shown). Minimum PVNS correspond with low rental rates and high discount rates.

For Building 5, the PVNS range from a minimum of -$2.9 million (compared to the 44 in (112 cm) exit stair) to a maximum of $12.7 million (compared to the 66 in (168 cm) exit stair). The mean PVNS are positive compared to the 56 in (142 cm) exit stair ($1.8 million) and for the 66 in (168 cm) exit stair ($3.6 million), but negative for the 44 in (112 cm) exit stair (-$0.2 million).

While the occupant evacuation elevator system (OEES) alternatives are generally cost-effective, at least when compared to the wider stair widths, the sensitivity analysis reveals combinations of the uncertainty parameters that are not (e.g., a negative PVNS). For the comparison with the 44 in (112 cm) exit stair this occurred in 59.2 % of the simulations (i.e., 5924 trials out of 10 000) (analysis not shown). The majority of variation in the PVNS measure (and SIR and AIRR, for that matter) can be explained by variation in the rental rate, discount rate, and Sky Lobby size. For the 44 in (112 cm) exit stair, 52 % of the variation in PVNS can be explained by changes in the size of the Sky Lobby, 27 % of the variation can be explained by variation in rental rate, and 20 % of the variation can be explained by the discount rate (analysis not shown). Minimum PVNS correspond with low rental rates, large Sky Lobby areas, and high discount rates.

For the comparison with the 56 in (142 cm) exit stair the OEES were not cost-effective in 13.0 % of the simulations (i.e., 1304 trials out of 10 000) (analysis not shown). The majority of variation in the PVNS measure (and SIR and AIRR, for that matter) can be explained by variation in the rental rate, discount rate, and Sky Lobby size. For the 56 in (142 cm) exit stair, 36 % of the variation in PVNS can be explained by changes in the rental rate, 36 % of the variation explained by the discount rate, and 28 % of the variation explained by variation in the size of the Sky Lobby (analysis not shown). Minimum PVNS correspond with low rental rates, large Sky Lobby areas, and high discount rates.

For the comparison with the 66 in (168 cm) exit stair the OEES were not cost-effective in 2.1 % of the simulations (i.e., 208 trials out of 10 000) (analysis not shown). The majority of variation in the PVNS measure (and SIR and AIRR, for that matter) can be explained by variation in the rental rate, discount rate, and Sky Lobby size. For the 66 in (168 cm) exit stair, 43 % of the variation explained by variation in the discount rate, 40 % of the variation in PVNS can be explained by changes in the rental rate, and 17 % of the variation explained by the size of the Sky Lobby (analysis not shown). Minimum PVNS correspond with low rental rates, large Sky Lobby areas, and high discount rates.

Table 4.28A Summary Statistics of the Economic Performance from the Monte Carlo Simulations of Occupant Evacuation Elevators compared to an Additional Exit Stair (PVNS in thousands of dollars)

Building	Floors	Economic Measure	Stair Width	Trials	Mean	Min	Median	Max	Std. Dev.
2	13	PVNS	44	10,000	623	(35)	608	1,386	779
2	13	PVNS	56	10,000	963	71	943	2,010	1,079
2	13	PVNS	66	10,000	1,279	166	1,253	2,595	1,361
3	28	PVNS	44	10,000	1,129	(274)	1,100	2,762	1,234
3	28	PVNS	56	10,000	1,865	(44)	1,822	4,102	1,649
3	28	PVNS	66	10,000	2,545	160	2,491	5,364	2,091
2	13	SIR	44	10,000	2.89	0.91	2.84	5.77	0.5137
2	13	SIR	56	10,000	3.67	1.18	3.61	7.20	0.6688
2	13	SIR	66	10,000	4.23	1.38	4.16	8.18	0.7902
3	28	SIR	44	10,000	2.28	0.72	2.25	4.47	0.4031
3	28	SIR	56	10,000	2.97	0.96	2.92	5.74	0.5117
3	28	SIR	66	10,000	3.50	1.15	3.44	6.68	0.6127
2	13	AIRR	44	10,000	0.09	0.04	0.09	0.14	0.0145
2	13	AIRR	56	10,000	0.10	0.06	0.10	0.15	0.0144
2	13	AIRR	66	10,000	0.11	0.06	0.10	0.15	0.0144
3	28	AIRR	44	10,000	0.08	0.03	0.08	0.13	0.0206
3	28	AIRR	56	10,000	0.09	0.05	0.09	0.14	0.0164
3	28	AIRR	66	10,000	0.10	0.05	0.10	0.14	0.0154

Table 4.28B Summary Statistics of the Economic Performance from the Monte Carlo Simulations of Occupant Evacuation Elevators compared to an Additional Exit Stair (PVNS in thousands of dollars)

Building	Floors	Economic Measure	Stair Width	Trials	Mean	Min	Median	Max	Std. Dev.
4	42	PVNS	44	10,000	1,377	(687)	1,337	3,792	780
4	42	PVNS	56	10,000	2,482	(341)	2,421	5,798	1,080
4	42	PVNS	66	10,000	3,502	(34)	3,425	7,691	1,362
5	75	PVNS	44	10,000	(151)	(2,880)	(301)	5,867	1,235
5	75	PVNS	56	10,000	1,822	(2,148)	1,639	9,410	1,651
5	75	PVNS	66	10,000	3,646	(1,599)	3,446	12,708	2,093
4	42	SIR	44	10,000	1.90	0.58	1.87	3.70	0.5139
4	42	SIR	56	10,000	2.53	0.80	2.49	4.84	0.6690
4	42	SIR	66	10,000	3.02	0.98	2.97	5.72	0.7905
5	75	SIR	44	10,000	0.95	0.08	0.90	3.05	0.4035
5	75	SIR	56	10,000	1.56	0.38	1.51	4.10	0.5121
5	75	SIR	66	10,000	2.06	0.56	2.01	4.94	0.6131
4	42	AIRR	44	10,000	0.07	0.03	0.07	0.12	0.0145
4	42	AIRR	56	10,000	0.08	0.04	0.08	0.13	0.0144
4	42	AIRR	66	10,000	0.09	0.05	0.09	0.14	0.0144
5	75	AIRR	44	10,000	0.04	(0.07)	0.04	0.10	0.0206
5	75	AIRR	56	10,000	0.06	0.00	0.06	0.12	0.0164
5	75	AIRR	66	10,000	0.07	0.02	0.07	0.13	0.0154

Note: The number of trials corresponds with the number of values used in the calculation of the statistics.

4.4.2.4 Break-Even Analysis

The discount rate and rental rate required for the investment in an OEES to break-even over its life-cycle is shown in Tables 4.29A and 4.29B. For example, the discount rate would need to exceed 13.5 % or the rental rate would need to be less than $17.14 per ft^2 ($184.43 per m^2) for the PVNS associated with Building 4 with 44 in (112 cm) stairs to be negative (not cost-effective). The discount rate would need to exceed 15.0 % or the rental rate would need to less than $15.79 per ft^2 ($169.96 per m^2) for the PVNS associated with Building 5 with 66 in (168 cm) stairs to be negative (not cost-effective).

Table 4.29A Discount Rate and Rental Rate Required for the Investment into Occupant Evacuation Elevators to Break-Even Over the Life-Cycle

Building	Floors	Stair Width (in)	Break-Even Discount Rate (%)	Break-Even Rental Rate in Dollars
2	13	44	20.9	11.63
2	13	56	26.7	9.17
2	13	66	30.8	7.89
3	28	44	16.4	14.36
3	28	56	21.6	11.16
3	28	66	25.5	9.47

Table 4.29B Discount Rate and Rental Rate Required for the Investment into Occupant Evacuation Elevators to Break-Even Over the Life-Cycle

Building	Floors	Stair Width (in)	Break-Even Discount Rate (%)	Break-Even Rental Rate in Dollars
4	42	44	13.5	17.14
4	42	56	18.3	13.20
4	42	66	22.1	11.10
5	75	44	5.4	29.74
5	75	56	10.9	20.11
5	75	66	15.0	15.79

4.4.3 Presentation and Analysis of Results

The results of the baseline analysis and the sensitivity analysis are summarized in Exhibits 4.1 (Building 2), 4.2 (Building 3), 4.3 (Building 4), and 4.4 (Building 5).

The summary format is based on ASTM Standard Guide E 2204; it was described in Section 3.3. Because the format is fairly compact, it was necessary to abbreviate some of the terms reported in the exhibits. As noted earlier, the term Base Case is used to represent the installation of an additional exit stair because the first cost for each of the three exit stair configurations was lower than the first cost for an occupant evacuation elevator system. The abbreviation BC refers to the Base Case (exit stair). The values in parentheses—(44), (56), and (66)—refer to the width of the additional exit stair. The abbreviation OEES refers to the occupant evacuation elevator system. The abbreviations BC(44), BC(56), BC(66), and OEES refer to the corresponding exit stair configurations and the occupant evacuation elevator system. For example, $LCC_{BC(44)}$ corresponds to the life-cycle cost of the 44 in (112 cm) wide exit stair. The abbreviations BC(44):OEES, BC(56):OEES, and BC(66):OEES are used to represent comparisons between a given exit stair width and the corresponding occupant evacuation elevator system. For example, $PVNS_{BC(66):OEES}$ corresponds to the present value net savings of the occupant evacuation elevator system vis-à-vis the 66 in (168 cm) wide exit stair.
This chapter presented tables summarizing cost data associated with four recently adopted egress-related requirements in the IBC. Two of the four requirements—

increased stair width for new sprinklered buildings and installation of a fire service access elevator in buildings over 120 ft (37 m) high—were analyzed from a life-cycle cost perspective. The two remaining requirements—installation of an additional exit stair in buildings over 420 ft (128 m) high and installation of occupant evacuation elevators—were analyzed via a two-step process. First, life-cycle costs were calculated and analyzed. Second, an economic analysis was performed to determine when the installation of occupant evacuation elevators was a cost-effective alternative to the required installation of an additional exit stair in buildings over 420 ft (128 m) high. The economic analysis included both a baseline analysis, where all inputs were held constant, and a sensitivity analysis, where key input variables were varied about their baseline values. The results of the baseline analysis demonstrate that occupant evacuation elevators are a cost-effective alternative to the installation of an additional exit stair for all four prototypical buildings over 120 ft (37 m) high, and not just for those prototypical buildings over 420 (128 m) high. Furthermore, these results are fairly robust, as demonstrated in the sensitivity analysis.

Exhibit 4.1 Summary of the Building 2 Cost-Effectiveness Analysis

1.a Significance of the Project:
Recent changes to the International Building Code (IBC) affect egress-related measures in buildings over 420 ft (128 m) high. GSA was interested in evaluating the changes for buildings over 120 ft (37 m) in height. Two such changes—an additional exit stair and permitting the use of occupant evacuation elevators as an alternative to the required addition of an exit stair—were the focus of an economic analysis. Information on the costs and specifications of alternative configurations for exit stairs and occupant evacuation elevators were compiled to support the economic analysis and to serve as a resource for building owners, fire protection engineers, and key construction industry stakeholders concerned about egress and life-safety issues in high rise buildings.

The economic analysis was commissioned by GSA in support of its objective to incorporate cost-effective fire protection and life safety systems that result in overall building safety that meets or exceeds the levels required by local building codes. The economic analysis was conducted in two phases. First a baseline analysis was performed holding all input variables at their most likely values. Second a sensitivity analysis employing Monte Carlo simulation was performed through which probabilistic levels of significance were calculated for the key measures of economic performance. The results of the economic analysis demonstrate that occupant evacuation elevators are a cost-effective alternative to the installation of an additional exit stair. Furthermore, these results are fairly robust, as demonstrated in the sensitivity analysis where key input variables were varied about their baseline values.

1.b Key Points:

- Recent changes to the IBC affect egress-related measures in buildings over 420 ft (128 m) high.
- GSA commissioned NIST to perform an economic analysis in support of its objective to incorporate cost-effective fire protection and life safety systems that result in overall building safety that meets or exceeds the levels required by local building codes.
- GSA was interested in evaluating the changes for buildings over 120 ft (37 m) in height.
- NIST compiled cost data on alternative configurations for exit stairs and occupant evacuation elevators to support the economic analysis and to serve as an information resource for building owners, fire protection engineers, and other key construction industry stakeholders concerned about egress and life-safety issues in high rise buildings.
- The results of the economic analysis demonstrate that occupant evacuation elevators are a cost-effective alternative to the installation of an additional exit stair.

2. Analysis Strategy: How Key Measures are Estimated

The following economic measures are calculated as present-value (PV) amounts:
(9) **Life-Cycle Costs** (LCC) for the Base Case (Additional Exit Stair of nominal width 44 in (112 cm), 56 in (142 cm), and 66 in (168 cm)) and for the Proposed Alternative (Occupant Evacuation Elevators), including all costs of installing and operating the two systems over the length of the study period. The selection criterion is lowest LCC.
(10) **Present Value Net Savings** (PVNS) that will result from selecting the lowest-LCC alternative. PVNS > 0 indicates an economically worthwhile project.

Additional measures:
(9) **Savings-to-Investment Ratio** (SIR), the ratio of savings from the lowest-LCC to the extra investment required to implement it. A ratio of SIR >1 indicates an economically worthwhile project.
(10) **Adjusted Internal Rate of Return** (AIRR), the annual return on investment over the study period. An AIRR > discount or hurdle rate indicates an economically worthwhile project.

Data and Assumptions:
- The Base Date is 2007.
- The alternative with the lower first cost (Additional Exit Stair) is designated the Base Case.
- The study period is 25 years and ends in 2031.
- The baseline value of the discount or hurdle rate is 2.7 % real.

Exhibit 4.1 Summary of Building 2 Cost-Effectiveness Analysis (Cont.)

3.a Calculation of Savings, Costs, and Additional Measures					
Results of Baseline Analysis (Savings and Costs in Thousands of Dollars)					
Economic Measure		Base Case (44)	OEES	Base Case (66)	OEES
Life-Cycle Cost (LCC)		$1,501	$551	$2,402	$551
Investment Cost		$114	$453	$48	$453
Delta Investment Cost		N/A	$339	N/A	$405
Non-Investment Cost		$1,387	$97	$2,354	$97
Savings		N/A	$1,290	N/A	$2,257
Present Value Net Savings (PVNS)		N/A	$950	N/A	$1,827
Savings-to-Investment Ratio (SIR)		N/A	3.80	N/A	5.57
AIRR		N/A	8.33 %	N/A	10.00 %
Results of Monte Carlo Simulation (Savings and Costs in Thousands of Dollars)					
Economic Measure	Statistical Measure				
	Minimum	Median	Maximum	Mean	Standard Deviation
$LCC_{BC(44)}$	495	1,134	1,949	1,150	260
$LCC_{BC(56)}$	601	1,469	2,578	1,490	353
$LCC_{BC(66)}$	695	1,779	3,163	5,832	440
LCC_{OEES}	447	527	601	527	26
$PVNS_{BC(44):OEES}$	-35	608	1,386	623	779
$PVNS_{BC(56):OEES}$	71	943	2,010	963	1,079
$PVNS_{BC(66):OEES}$	166	1,253	2,595	1,279	1,361

3.b Key Results		3.c Traceability
*LCC (Thousands of Dollars)		Life-cycle costs and supplementary measures were calculated according to ASTM standards E 917, E 964, E 1057, and E 1074. Treatment of uncertainty and measures of project risk were calculated according to ASTM standards E 1369 and E 1946. Section 3008 of the 2009 edition of the International Building Code specifies the criteria that passenger elevators must meet to be used for evacuation purposes.
Base Case (44)	$1,501	
Base Case (56)	$1,968	
Base Case (66)	$2,402	
OEES	$551	
*PVNS (Thousands of Dollars)		
BC(44):OEES	$951	
BC(56):OEES	$1,418	
BC(66):OEES	$1,852	
*SIR		
BC(44):OEES	3.80	
BC(56):OEES	4.82	
BC(66):OEES	5.57	
*AIRR		
BC(44):OEES	8.33 %	
BC(56):OEES	9.37 %	
BC(66):OEES	10.00 %	

Exhibit 4.2 Summary of the Building 3 Cost-Effectiveness Analysis

| **1.a Significance of the Project:**
Recent changes to the International Building Code (IBC) affect egress-related measures in buildings over 420 ft (128 m) high. GSA was interested in evaluating the changes for buildings over 120 ft (37 m) in height. Two such changes—an additional exit stair and permitting the use of occupant evacuation elevators as an alternative to the required addition of an exit stair—were the focus of an economic analysis. Information on the costs and specifications of alternative configurations for exit stairs and occupant evacuation elevators were compiled to support the economic analysis and to serve as a resource for building owners, fire protection engineers, and key construction industry stakeholders concerned about egress and life-safety issues in high rise buildings.

The economic analysis was commissioned by GSA in support of its objective to incorporate cost-effective fire protection and life safety systems that result in overall building safety that meets or exceeds the levels required by local building codes. The economic analysis was conducted in two phases. First a baseline analysis was performed holding all input variables at their most likely values. Second a sensitivity analysis employing Monte Carlo simulation was performed through which probabilistic levels of significance were calculated for the key measures of economic performance. The results of the economic analysis demonstrate that occupant evacuation elevators are a cost-effective alternative to the installation of an additional exit stair. Furthermore, these results are fairly robust, as demonstrated in the sensitivity analysis where key input variables were varied about their baseline values. | **1.b Key Points:**
• Recent changes to the IBC affect egress-related measures in buildings over 420 ft (128 m) high.
• GSA commissioned NIST to perform an economic analysis in support of its objective to incorporate cost-effective fire protection and life safety systems that result in overall building safety that meets or exceeds the levels required by local building codes.
• GSA was interested in evaluating the changes for buildings over 120 ft (37 m) in height.
• NIST compiled cost data on alternative configurations for exit stairs and occupant evacuation elevators to support the economic analysis and to serve as an information resource for building owners, fire protection engineers, and other key construction industry stakeholders concerned about egress and life-safety issues in high rise buildings.
• The results of the economic analysis demonstrate that occupant evacuation elevators are a cost-effective alternative to the installation of an additional exit stair. |

2. Analysis Strategy: How Key Measures are Estimated

The following economic measures are calculated as present-value (PV) amounts:

(11) **Life-Cycle Costs** (LCC) for the Base Case (Additional Exit Stair of nominal width 44 in (112 cm), 56 in (142 cm), and 66 in (168 cm)) and for the Proposed Alternative (Occupant Evacuation Elevators), including all costs of installing and operating the two systems over the length of the study period. The selection criterion is lowest LCC.

(12) **Present Value Net Savings** (PVNS) that will result from selecting the lowest-LCC alternative. PVNS > 0 indicates an economically worthwhile project.

Additional measures:

(11) **Savings-to-Investment Ratio** (SIR), the ratio of savings from the lowest-LCC to the extra investment required to implement it. A ratio of SIR >1 indicates an economically worthwhile project.

(12) **Adjusted Internal Rate of Return** (AIRR), the annual return on investment over the study period. An AIRR > discount or hurdle rate indicates an economically worthwhile project.

Data and Assumptions:
- The Base Date is 2007.
- The alternative with the lower first cost (Additional Exit Stair) is designated the Base Case.
- The study period is 25 years and ends in 2031.
- The baseline value of the discount or hurdle rate is 2.7 % real.

Exhibit 4.2 Summary of Building 3 Cost-Effectiveness Analysis (Cont.)

3.a Calculation of Savings, Costs, and Additional Measures					
Results of Baseline Analysis (Savings and Costs in Thousands of Dollars)					
Economic Measure	Base Case (44)	OEES	Base Case (66)	OEES	
Life-Cycle Cost (LCC)	$3,229	$1,403	$5,173	$1,403	
Investment Cost	241	$1,143	$103	$1,143	
Delta Investment Cost	N/A	$902	N/A	$1040	
Non-Investment Cost	$2,988	$259	$5,070	$259	
Savings	N/A	$2729	N/A	$4811	
Present Value Net Savings (PVNS)	N/A	$1,827	N/A	$3,770	
Savings-to-Investment Ratio (SIR)	N/A	3.02	N/A	4.62	
AIRR	N/A	7.35 %	N/A	9.19 %	
Results of Monte Carlo Simulation (Savings and Costs in Thousands of Dollars)					
Economic Measure	Statistical Measure				
	Minimum	Median	Maximum	Mean	Standard Deviation
$LCC_{BC(44)}$	1,063	2,438	4,195	2,472	559
$LCC_{BC(56)}$	1,293	3,162	5,550	3,208	760
$LCC_{BC(66)}$	1,497	3,830	6,812	3,888	949
LCC_{OEES}	1,152	1,344	1,522	1,343	62
$PVNS_{BC(44):OEES}$	-274	1,100	2,762	1,129	1,234
$PVNS_{BC(56):OEES}$	-44	1,822	4,102	1,865	1,649
$PVNS_{BC(66):OEES}$	160	2,491	5,364	2,545	2,091

3.b Key Results		3.c Traceability
*LCC (Thousands of Dollars)		Life-cycle costs and supplementary measures were calculated according to ASTM standards E 917, E 964, E 1057, and E 1074. Treatment of uncertainty and measures of project risk were calculated according to ASTM standards E 1369 and E 1946. Section 3008 of the 2009 edition of the International Building Code specifies the criteria that passenger elevators must meet to be used for evacuation purposes.
Base Case (44)	$3,229	
Base Case (56)	$4,237	
Base Case (66)	$5,173	
OEES	$1,403	
*PVNS (Thousands of Dollars)		
BC(44):OEES	$1,827	
BC(56):OEES	$2,835	
BC(66):OEES	$3,770	
*SIR		
BC(44):OEES	3.02	
BC(56):OEES	3.93	
BC(66):OEES	4.62	
*AIRR		
BC(44):OEES	7.35 %	
BC(56):OEES	8.48 %	
BC(66):OEES	9.19 %	

Exhibit 4.3 Summary of the Building 4 Cost-Effectiveness Analysis

1.a Significance of the Project:	1.b Key Points:
Recent changes to the International Building Code (IBC) affect egress-related measures in buildings over 420 ft (128 m) high. Two such changes—an additional exit stair and permitting the use of occupant evacuation elevators as an alternative to the required addition of an exit stair—were the focus of an economic analysis. Information on the costs and specifications of alternative configurations for exit stairs and occupant evacuation elevators were compiled to support the economic analysis and to serve as a resource for building owners, fire protection engineers, and key construction industry stakeholders concerned about egress and life-safety issues in high rise buildings. The economic analysis was commissioned by GSA in support of its objective to incorporate cost-effective fire protection and life safety systems that result in overall building safety that meets or exceeds the levels required by local building codes. The economic analysis was conducted in two phases. First a baseline analysis was performed holding all input variables at their most likely values. Second a sensitivity analysis employing Monte Carlo simulation was performed through which probabilistic levels of significance were calculated for the key measures of economic performance. The results of the economic analysis demonstrate that occupant evacuation elevators are a cost-effective alternative to the installation of an additional exit stair. Furthermore, these results are fairly robust, as demonstrated in the sensitivity analysis where key input variables were varied about their baseline values.	• Recent changes to the IBC affect egress-related measures in buildings over 420 ft (128 m) high. • GSA commissioned NIST to perform an economic analysis in support of its objective to incorporate cost-effective fire protection and life safety systems that result in overall building safety that meets or exceeds the levels required by local building codes. • NIST compiled cost data on alternative configurations for exit stairs and occupant evacuation elevators to support the economic analysis and to serve as an information resource for building owners, fire protection engineers, and other key construction industry stakeholders concerned about egress and life-safety issues in high rise buildings. • The results of the economic analysis demonstrate that occupant evacuation elevators are a cost-effective alternative to the installation of an additional exit stair.

2. Analysis Strategy: How Key Measures are Estimated

The following economic measures are calculated as present-value (PV) amounts:

(13) **Life-Cycle Costs** (LCC) for the Base Case (Additional Exit Stair of nominal width 44 in (112 cm), 56 in (142 cm), and 66 in (168 cm)) and for the Proposed Alternative (Occupant Evacuation Elevators), including all costs of installing and operating the two systems over the length of the study period. The selection criterion is lowest LCC.

(14) **Present Value Net Savings** (PVNS) that will result from selecting the lowest-LCC alternative. PVNS > 0 indicates an economically worthwhile project.

Additional measures:

(13) **Savings-to-Investment Ratio** (SIR), the ratio of savings from the lowest-LCC to the extra investment required to implement it. A ratio of SIR >1 indicates an economically worthwhile project.

(14) **Adjusted Internal Rate of Return** (AIRR), the annual return on investment over the study period. An AIRR > discount or hurdle rate indicates an economically worthwhile project.

Data and Assumptions:
- The Base Date is 2007.
- The alternative with the lower first cost (Additional Exit Stair) is designated the Base Case.
- The study period is 25 years and ends in 2031.
- The baseline value of the discount or hurdle rate is 2.7 % real.

Exhibit 4.3 Summary of Building 4 Cost-Effectiveness Analysis (Cont.)

3.a Calculation of Savings, Costs, and Additional Measures					
Results of Baseline Analysis (Savings and Costs in Thousands of Dollars)					
Economic Measure	Base Case (44)	OEES	Base Case (66)	OEES	
Life-Cycle Cost (LCC)	$4,842	$2,440	$7,759	$2,440	
Investment Cost	$360	$1,921	$154	$1,921	
Delta Investment Cost	N/A	$1,561	N/A	$1,767	
Non-Investment Cost	$4,482	$519	$7,605	$519	
Savings	N/A	$3,963	N/A	$7,086	
Present Value Net Savings (PVNS)	N/A	$2,402	N/A	$5,319	
Savings-to-Investment Ratio (SIR)	N/A	2.54	N/A	4.01	
AIRR	N/A	6.60 %	N/A	8.57 %	
Results of Monte Carlo Simulation (Savings and Costs in Thousands of Dollars)					
Economic Measure	Statistical Measure				
	Minimum	Median	Maximum	Mean	Standard Deviation
$LCC_{BC(44)}$	1,593	3,655	6,291	3,707	839
$LCC_{BC(56)}$	1,938	4,742	8,324	4,811	1,140
$LCC_{BC(66)}$	2,245	5,745	10,218	5,832	1,423
LCC_{OEES}	2,004	2,330	2,644	2,330	109
$PVNS_{BC(44):OEES}$	-687	1,337	3,792	1,377	780
$PVNS_{BC(56):OEES}$	-341	2,421	5,798	2,482	1,080
$PVNS_{BC(66):OEES}$	-34	3,425	7,691	3,502	1,362

3.b Key Results		3.c Traceability
*LCC (Thousands of Dollars)		Life-cycle costs and supplementary measures were calculated according to ASTM standards E 917, E 964, E 1057, and E 1074. Treatment of uncertainty and measures of project risk were calculated according to ASTM standards E 1369 and E 1946. Section 3008 of the 2009 edition of the International Building Code specifies the criteria that passenger elevators must meet to be used for evacuation purposes.
Base Case (44)	$4,842	
Base Case (56)	$6,355	
Base Case (66)	$7,759	
OEES	$2,440	
*PVNS (Thousands of Dollars)		
BC(44):OEES	$2,402	
BC(56):OEES	$3,915	
BC(66):OEES	$5,319	
*SIR		
BC(44):OEES	2.54	
BC(56):OEES	3.36	
BC(66):OEES	4.01	
*AIRR		
BC(44):OEES	6.60 %	
BC(56):OEES	7.80 %	
BC(66):OEES	8.57 %	

Exhibit 4.4 Summary of the Building 5 Cost-Effectiveness Analysis

1.a Significance of the Project:	1.b Key Points:
Recent changes to the International Building Code (IBC) affect egress-related measures in buildings over 420 ft (128 m) high. Two such changes—an additional exit stair and permitting the use of occupant evacuation elevators as an alternative to the required addition of an exit stair—were the focus of an economic analysis. Information on the costs and specifications of alternative configurations for exit stairs and occupant evacuation elevators were compiled to support the economic analysis and to serve as a resource for building owners, fire protection engineers, and key construction industry stakeholders concerned about egress and life-safety issues in high rise buildings. The economic analysis was commissioned by GSA in support of its objective to incorporate cost-effective fire protection and life safety systems that result in overall building safety that meets or exceeds the levels required by local building codes. The economic analysis was conducted in two phases. First a baseline analysis was performed holding all input variables at their most likely values. Second a sensitivity analysis employing Monte Carlo simulation was performed through which probabilistic levels of significance were calculated for the key measures of economic performance. The results of the economic analysis demonstrate that occupant evacuation elevators are a cost-effective alternative to the installation of an additional exit stair. Furthermore, these results are fairly robust, as demonstrated in the sensitivity analysis where key input variables were varied about their baseline values.	• Recent changes to the IBC affect egress-related measures in buildings over 420 ft (128 m) high. • GSA commissioned NIST to perform an economic analysis in support of its objective to incorporate cost-effective fire protection and life safety systems that result in overall building safety that meets or exceeds the levels required by local building codes. • NIST compiled cost data on alternative configurations for exit stairs and occupant evacuation elevators to support the economic analysis and to serve as an information resource for building owners, fire protection engineers, and other key construction industry stakeholders concerned about egress and life-safety issues in high rise buildings. • The results of the economic analysis demonstrate that occupant evacuation elevators are a cost-effective alternative to the installation of an additional exit stair.

2. Analysis Strategy: How Key Measures are Estimated

The following economic measures are calculated as present-value (PV) amounts:

(15) **Life-Cycle Costs** (LCC) for the Base Case (Additional Exit Stair of nominal width 44 in (112 cm), 56 in (142 cm), and 66 in (168 cm)) and for the Proposed Alternative (Occupant Evacuation Elevators), including all costs of installing and operating the two systems over the length of the study period. The selection criterion is lowest LCC.

(16) **Present Value Net Savings** (PVNS) that will result from selecting the lowest-LCC alternative. PVNS > 0 indicates an economically worthwhile project.

Additional measures:

(15) **Savings-to-Investment Ratio** (SIR), the ratio of savings from the lowest-LCC to the extra investment required to implement it. A ratio of SIR >1 indicates an economically worthwhile project.

(16) **Adjusted Internal Rate of Return** (AIRR), the annual return on investment over the study period. An AIRR > discount or hurdle rate indicates an economically worthwhile project.

Data and Assumptions:
- The Base Date is 2007.
- The alternative with the lower first cost (Additional Exit Stair) is designated the Base Case.
- The study period is 25 years and ends in 2031.
- The baseline value of the discount or hurdle rate is 2.7 % real.

Exhibit 4.4 Summary of Building 5 Cost-Effectiveness Analysis (Cont.)

3.a Calculation of Savings, Costs, and Additional Measures					
Results of Baseline Analysis (Savings and Costs in Thousands of Dollars)					
Economic Measure		Base Case (44)	OEES	Base Case (66)	OEES
Life-Cycle Cost (LCC)		$8,644	$7,604	$13,855	$7,604
Investment Cost		$640	$3,761	$274	$3,761
Delta Investment Cost		N/A	$3,121	N/A	$3,487
Non-Investment Cost		$8,004	$3,843	$13,581	$3,843
Savings		N/A	$4,161	N/A	$9,738
Present Value Net Savings (PVNS)		N/A	$1,040	N/A	$6,251
Savings-to-Investment Ratio (SIR)		N/A	1.33	N/A	2.79
AIRR		N/A	3.89 %	N/A	7.01 %
Results of Monte Carlo Simulation (Savings and Costs in Thousands of Dollars)					
Economic Measure	Statistical Measure				
	Minimum	Median	Maximum	Mean	Standard Deviation
$LCC_{BC(44)}$	2,841	6,525	11,232	6,616	1,498
$LCC_{BC(56)}$	3,459	8,466	14,864	8,590	2,036
$LCC_{BC(66)}$	4,009	10,258	18,245	10,414	2,542
LCC_{OEES}	4,271	6,641	11,408	6,768	1,114
$PVNS_{BC(44):OEES}$	-2,880	-301	5,876	-151	1,235
$PVNS_{BC(56):OEES}$	-2,148	1,639	9,410	1,822	1,651
$PVNS_{BC(66):OEES}$	-1,599	3,446	12,708	3,646	2,093

3.b Key Results

*LCC (Thousands of Dollars)

Base Case (44)	$8,644
Base Case (56)	$11,347
Base Case (66)	$13,855
OEES	$7,604

*PVNS (Thousands of Dollars)

BC(44):OEES	$1,040
BC(56):OEES	$3,743
BC(66):OEES	$6,251

*SIR

BC(44):OEES	1.33
BC(56):OEES	2.14
BC(66):OEES	2.79

*AIRR

BC(44):OEES	3.89 %
BC(56):OEES	5.87 %
BC(66):OEES	7.01 %

3.c Traceability

Life-cycle costs and supplementary measures were calculated according to ASTM standards E 917, E 964, E 1057, and E 1074. Treatment of uncertainty and measures of project risk were calculated according to ASTM standards E 1369 and E 1946.

Section 3008 of the 2009 edition of the International Building Code specifies the criteria that passenger elevators must meet to be used for evacuation purposes.

5 Summary and Recommendations for Further Research

5.1 Summary

Fire protection measures are needed to maintain the safety and integrity of the Nation's building stock and to limit loss of life and property when building fires do occur. Statistics published by the National Fire Protection Association (NFPA) demonstrate that fire protection is a major investment cost in building construction. Therefore, ways to reduce the costs of fire protection while ensuring safety are of interest to building owners, fire protection engineers, and other construction industry stakeholders. Fire protection measures include, but are not limited to, building safety features concerned with extinguishment (e.g., sprinklers), containment (e.g., compartmentation), passive resistance (e.g., fire resistive materials), detection and alarm (e.g., smoke detectors), and egress (e.g., exit stairs). Although all fire protection measures have important economic implications, both in terms of first costs and future costs associated with operations and maintenance, the focus of this report is on egress-related measures in new building construction.

Recent changes in the International Building Code (IBC) have set the stage for a analyzing the costs of several key egress-related requirements. The U.S. General Services Administration (GSA) commissioned this study to conduct an economic analysis of the use of elevators and exit stairs for occupant evacuation and fire service access. The goal of this report is to produce economic analyses of cost data suitable for evaluating improved egress system designs that promote efficient and timely egress of occupants, including those with disabilities, and that facilitate more efficient fire department operations. This report tabulates cost data for selected egress-related requirements in five prototypical buildings specified by GSA. The five prototypical buildings range in height from a 5-floor, mid-rise building to a 75-floor, high-rise building. Cost data are tabulated in a format that facilitates life-cycle cost analyses of selected egress-related requirements. Incremental costs are also tabulated to help assess the implications of changing one or more design parameters.

The results of the economic analysis for the two prototypical buildings over 420 ft (128 m) high, demonstrate that: (1) an additional exit stair is a cost-effective alternative to the installation of occupant evacuation elevators on a first-cost basis; and (2) occupant evacuation elevators are a cost-effective alternative to the installation of an additional exit stair on a life-cycle cost basis when rental rates are high and discount rates are low.

The specifications, assumptions, and cost estimating relationships of the occupant evacuation elevators were developed in consultation with industry experts. The elevator configuration and related cost elements, described in Appendix A, are representative of one design possibility. Others may exist.

5.2 Recommendations for Further Research

The background work for this report uncovered an additional area of research that would be of value to government agencies, private-sector organizations, building owners and managers, and fire protection engineers concerned with the relative costs of alternative fire protection measures. This area of research is concerned with the development of a revised procedure for estimating the national costs of fire protection for building construction.

The current procedure for estimating the national cost of fire protection in buildings needs to be revised to incorporate changes in codes, standards, practices, and technologies that have been adopted over the last 20 to 30 years. Once completed, the revised procedure will provide a comprehensive, science-based approach for estimating the national cost of fire protection in buildings. The approach will be based on cost estimating relationships applied to the value of construction put in place statistics produced by the U.S. Census Bureau in its C30 Report. Thus, the revised procedure will enable annual estimates for the national cost of fire protection to be produced. An added advantage of the proposed approach is that it will enable not only the overall national cost of fire protection in buildings to be estimated but also the national cost for each of the key fire protection strategies (e.g., extinguishment, containment, passive resistance, detection and alarm, and egress) to be estimated. No such information is currently available and represents a serious gap in understanding how the costs of fire protection are distributed among fire protection strategies. Ideally, the approach will include an "assembly-level" summary of fire protection costs for each fire protection strategy (e.g., occupant evacuation elevators) that will facilitate analyses of key cost drivers for specific fire safety measures (e.g., water protection for occupant evacuation elevators) covered in an assembly.

References

Apostolou, J.J., D.L. Bowers, and C.M. Sullivan. 1978. *The Nation's Annual Expenditure for the Prevention and Control of Fire.* Project Report. Worcester, MA: Worcester Polytechnic Institute.

ASTM International. "Standard Guide for Selecting Economic Methods for Evaluating Investments in Buildings and Building Systems," E 1185, *Annual Book of ASTM Standards: 2008*, Vol. 04.11. West Conshohocken, PA: ASTM International.

ASTM International. "Standard Guide for Selecting Techniques for Treating Uncertainty and Risk in the Economic Evaluation of Buildings and Building Systems," E 1369, *Annual Book of ASTM Standards: 2008*, Vol. 04.11. West Conshohocken, PA: ASTM International.

ASTM International. "Standard Guide for Summarizing Economic Impacts of Building Related Projects," E 2204, *Annual Book of ASTM Standards: 2008*, Vol. 04.12. West Conshohocken, PA: ASTM International.

ASTM International. "Standard Practice for Applying Analytical Hierarchy Process (AHP) to Multiattribute Decision Analysis of Investments Related to Buildings and Building Systems," E 1765, *Annual Book of ASTM Standards: 2008*, Vol. 04.12. West Conshohocken, PA: ASTM International.

ASTM International. "Standard Practice for Measuring Benefit-to-Cost and Savings-to-Investment Ratios for Investments in Buildings and Building Systems," E 964, *Annual Book of ASTM Standards: 2008*, Vol. 04.11. West Conshohocken, PA: ASTM International.

ASTM International. "Standard Practice for Measuring Cost Risk of Buildings and Building Systems," E 1946, *Annual Book of ASTM Standards: 2008*, Vol. 04.12. West Conshohocken, PA: ASTM International.

ASTM International. "Standard Practice for Measuring Internal Rate of Return and Adjusted Internal Rate of Return for Investments in Buildings and Building Systems," E 1057, *Annual Book of ASTM Standards: 2008*, Vol. 04.11. West Conshohocken, PA: ASTM International.

ASTM International. "Standard Practice for Measuring Life-Cycle Costs of Buildings and Building Systems," E 917, *Annual Book of ASTM Standards: 2008*, Vol. 4.11. West Conshohocken, PA: ASTM International.

ASTM International. "Standard Practice for Measuring Net Benefits and Net Savings for Investments in Buildings and Building Systems," E 1074, *Annual Book of ASTM Standards: 2008*, Vol. 04.11. West Conshohocken, PA: ASTM International.

Averill, J.D., and W. Song. 2007. *Accounting for Emergency Response in Building Evacuation: Modeling Differential Egress Capacity Solutions.* NISTIR 7425. Gaithersburg, MD: National Institute of Standards and Technology.

BOMA (Building Owners and Managers Association) International, Experience Exchange Report. Washington, DC: BOMA International. CD-ROM.

Bureau of Economic Analysis. "Gross-Domestic-Product-(GDP)-by-Industry Data." *Industry Economic Accounts* (Washington, DC: Bureau of Economic Analysis), http://www.bea.gov/bea/dn2/gdpbyind_data.htm (accessed December 2009).

Chapman, R.E., and Fuller, S.K. 1996. *Benefits and Costs of Research: Two Case Studies in Building Technology.* NISTIR 5840. Gaithersburg, MD: National Institute of Standards and Technology.

Crystal Ball. 2007. *Crystal Ball 7.3 User Manual.* Denver, CO: Decisioneering, Inc.

Hall, J.R. 2008. *The Total Cost of Fire in the United States.* Quincy, MA: National Fire Protection Association.

Harris, Carl M. 1984. *Issues in Sensitivity and Statistical Analysis of Large-Scale, Computer-Based Models.* NBS GCR 84-466. Gaithersburg, MD: National Bureau of Standards.

International Code Council, Inc. 2009. *International Building Code.* Washington, DC: International Code Council, Inc.

McKay, M. C., W. H. Conover, and R.J. Beckman. 1979. "A Comparison of Three Methods for Selecting Values of Input Variables in the Analysis of Output from a Computer Code." *Technometrics* (Vol. 21): pp. 239-245.

Meade, W.P. 1991. *A First Pass at Computing Fire Safety in a Modern Society.* NIST-GCR-91-592. Gaithersburg, MD: National Institute of Standards and Technology.

National Institute of Standards and Technology. "Safer Buildings Are Goal of New Code Changes Based on Recommendations from NIST World Trade Center Investigation" TechBeat: October 1, 2008. http://www.nist.gov/public_affairs/releases/wtc_100108.html (accessed December 2008).

Office of Management and Budget. "Circular A-94: Guidelines and Discount Rates for Benefit-Cost Analysis of Federal Programs."

Orszag, P.R. "Memorandum for the Heads of Departments and Agencies: 2010 Discount Rates for OMB Circular No. A-94," Office of Management and Budget, December 8, 2009.

Reed Construction Data, Inc. 2008. *RSMeans Construction Cost Data.* 66th Edition. Kingston, MA: Reed Construction Data, Inc.

Reed Construction Data, Inc. 2009. *RSMeans Costworks.* Kingston, MA: Reed Construction Data, Inc. CD-ROM.

Templer, J.A. Stair Shape and Human Movement, Ph.D. dissertation. New York, NY: Columbia University, (1974).

United States Census Bureau: Manufacturing and Construction Division. "Annual Value of Construction Put in Place." *Current Construction Report (CCR) C30 (*Washington, DC: United States Census Bureau, August 1, 2009), http://www.census.gov/const/C30/total.pdf (accessed December 2009).

United States Department of Energy: Energy Information Administration. "Commercial Buildings Energy Consumption Survey." (Washington DC: United States Department of Energy, September 2008), http://www.eia.doe.gov/emeu/cbecs/cbecs2003/detailed_tables_2003/2003set1/2003excel/a1.xls (accessed December 2008).

United States Department of Energy: Energy Information Administration. "Residential Energy Consumption Survey." (Washington DC: United States Department of Energy, April 2008), http://www.eia.doe.gov/emeu/recs/recs2005/hc2005_tables/detailed_tables2005.html (accessed December 2008).

United States General Services Administration. *Exit Stair Cost Analysis.* (Washington, DC: United States General Services Administration, July 12, 2007).

United States General Services Administration, Public Building Service. *Facilities Standards for the Public Buildings Service*, PBS-100. (Washington, DC: U.S. General Services Administration, March 2005.

Whitestone Research. 2008. *The Whitestone Building Maintenance and Repair Cost Reference: 2008-2009.* 13th Edition. Santa Barbara, CA: Whitestone Research.

Appendix A Summary of Cost Estimates for Converting Passenger Elevators to Occupant Evacuation Elevators

This appendix contains the cost estimates for converting standard passenger elevators to occupant evacuation elevators in Buildings 2 through 5. These costs include water protection costs, signage, lobby status indicators, two-way communication systems, protection of wiring, lobby enclosures, and annual maintenance costs. Building 5 has additional costs for a sky lobby. The table for each building is broken up into sections. Section I describes characteristics of the building, section II describes the elevator requirements for the building, and the final two sections include the total initial capital costs and the total annually recurring costs. The remaining sections describe individual costs. When applicable, metric units have been provided in a column on the right side of the page.

Building 5 has two zones of elevators. The low-zone elevators service floors 1 through 38 while the high-zone elevators service floors 39 through 75. Shuttle elevators go to a sky lobby where individuals can transfer to high-zone elevators. Dimensions of the sky lobby are provided along with the cost estimates. This lobby must be protected; these costs are described in section VII, Sky Lobby Increment. The lobby is required to be large enough that at 25 % occupant load there is 3 ft^2 per person in addition to 10 ft^2 for every 50 people served by the shuttle elevators. Also notice that within each zone there are elevators that each service different floors. Buildings 3 and 4 also have elevators that each service differing floors, but do not have a sky lobby.

Some cost items are present on every floor that an elevator travels through while others are only present on floors that an elevator stops. Hoistways/landings are present on every floor that an elevator travels through while entrances and lobbies occur only on the floors that an elevator stops. Water protection costs include the installation of a linear drainage channel in front of each elevator hoistway and associated elevator landing. These occur at every floor an elevator travels through. Water protection costs also include installation of interlocking elements on elevator frames and door panels to provide seal for the elevator entrance and installation of door interlocks and wiring devices on the hoistway doors. These occur at each floor that an elevator stops.

Summary of Cost Estimates for Converting Passenger Elevators to Occupant Evacuation Elevators: Building 2

I. Characteristics
    a. Number of Floors: 13
    b. Building Height: 156 ft   48 m
    c. Per Floor Area: 25,000 ft^2 2,323 m^2

II. Passenger Elevator Requirements
    a. Number of Elevators: 6
    b. Elevator Weight: 3,500 lbs   1,588 kg
    c. Travel Speed: 700 ft/min   3.6 m·s^{-1}
    d. Number of Floors Served: 13
    e. Entrances: 78

III. Water Protection Costs
    a. Installation of a linear drainage channel in front of each elevator hoistway and associated elevator landing:
        i. Cost per hoistway/landing $500
        ii. Number of hoistways/landings (6 elevators traveling 13 floors)   x   78
        iii. Total $39,000
    b. Installation of interlocking elements on elevator frames and door panels to provide seal for the elevator entrances:
        i. Cost per entrance $250
        ii. Number of entrances (6 elevators serving 13 floors)   x   78
        iii. Total $19,500
    c. Installation of weatherproof door interlocks and wiring devices on the hoistway doors:
        i. Cost per entrance $500
        ii. Number of entrances (6 elevators serving 13 floors)   x   78
        iii. Total $39,000
    d. Total water protection costs   **$97,500**

IV. Signage, Lobby Status Indicator, and Two-Way Communication System
    a. Cost per elevator lobby $4,500
    b. Number of lobbies (above ground floor)   x   12
    c. Total   **$54,000**

V. Protection of Wiring or Cables
    a. Cost per elevator $300
    b. Number of elevators   x   6
    c. Total   **$1,800**

VI. Lobby Enclosure Costs

a. Installation of two hollow-core steel fire doors and frame system with vision panels and automatic closing devices:
i. Cost per set		$5,250
ii. Set per lobby		4
iii. Number of lobbies (above ground floor)	x	12
iv. Total		$252,000

b. Sealing of interstitial spaces:
i. Cost per passenger elevator lobby		$4,000 ($25 per linear foot)
ii. Number of lobbies (above ground floor)	x	12
iii. Total		$48,000

c. Total lobby enclosure costs   **$300,000**

VII. Maintenance Costs

a. Cost per elevator (above and beyond a passenger elevator)		$900
b. Number of passenger elevators	x	6
c. Total		**$5,400**

VIII. Initial Capital Costs of Converting passenger Elevators to Occupant Evacuation Elevators

a. Water protection	$97,500
b. Signage, lobby status indicator, and two-way communication	$54,000
c. Protection of wiring/cables	$1,800
d. Lobby enclosure	$300,000
e. Total	**$453,300**

IX. Annually Recurring Costs of Converting passenger Elevators to Occupant Evacuation Eleva

a. Maintenance	$5,400
b. Total	**$5,400**

Summary of Cost Estimates for Converting Passenger Elevators to Occupant Evacuation Elevators: Building 3

I. Characteristics
    a. Number of Floors: 28
    b. Building Height: 336 ft   102 m
    c. Per Floor Area: 30,000 ft^2 2,787 m^2

II. Passenger Elevator Requirements
    a. Low Rise Elevators:
        i. Number of Elevators   8
        ii. Elevator Weight: 3,500 lbs   1,588 kg
        iii. Travel Speed   700 ft/min   3.6 m·s^{-1}
        iv. Number of Floors Served   14
        v. Floors Serviced by Elevator   1-14
        vi. Number of Entrances (8 elevators for 14 floors)   112
    b. Mid Low Rise Elevators:
        i. Number of Elevators   8
        ii. Elevator Weight: 3,500 lbs   1,588 kg
        iii. Travel Speed   1,000 ft/min   3.6 m·s^{-1}
        iv. Number of Floors Served   15
        v. Floors Serviced by Elevator   1, 15-28
        vi. Number of Entrances (8 elevators for 15 floors)   120

III. Water Protection Costs
    a. Installation of a linear drainage channel in front of each elevator hoistway and associated elevator landing:
        i. Cost per hoistway/landing $500
        ii. Number of hoistways/landings (8 low rise elevators traveling 14 floors and 8 mid low elevators traveling 28 floors)   x   336
        iii. Total $168,000
    b. Installation of interlocking elements on elevator frames and door panels to provide seal for the elevator entrances:
        i. Cost per entrance $250
        ii. Number of entrances (8 low rise elevators serving 14 floors and 8 mid low elevators serving 15 floors)   x   232
        iii. Total $58,000
    c. Installation of weatherproof door interlocks and wiring devices on the hoistway doors:
        i. Cost per entrance $500
        ii. Number of entrances (8 low rise elevators serving 14 floors and 8 mid low elevators serving 15 floors)   x   232
        iii. Total $116,000

d. Total water protection costs		$342,000

IV. Signage, Lobby Status Indicator, and Two-Way Communication System
a. Cost per elevator lobby		$4,500
b. Number of lobbies (above ground floor)	x	27
c. Total		**$121,500**

V. Protection of Wiring or Cables
a. Cost per elevator		$300
b. Number of elevators	x	16
c. Total		**$4,800**

VI. Lobby Enclosure Costs
a. Installation of two hollow-core steel fire doors and frame system with vision panels and automatic closing devices:

i. Cost per set		$5,250
ii. Sets per lobby		4
iii. Number of lobbies (above ground floor)	x	27
iv. Total		$567,000

b. Sealing of interstitial spaces:

i. Cost per passenger elevator lobby		$4,000 ($25 per linear foot)
ii. Number of lobbies (above ground floor)	x	27
iii. Total		$108,000

c. Total lobby enclosure costs		**$675,000**

VII. Maintenance Costs
a. Cost per elevator (above and beyond a passenger elevator)		$900
b. Number of passenger elevators	x	16
c. Total		**$14,400**

VIII. Initial Capital Costs of Converting passenger Elevators to Occupant Evacuation Elevators
a. Water protection	$342,000
b. Signage, lobby status indicator, and two-way communication	$121,500
c. Protection of wiring/cables	$4,800
d. Lobby enclosure	$675,000
e. Total	**$1,143,300**

IX. Annually Recurring Costs of Converting passenger Elevators to Occupant Evacuation Elevato
a. Maintenance	$14,400
b. Total	**$14,400**

Summary of Cost Estimates for Converting Passenger Elevators to Occupant Evacuation Elevators: Building 4

I. Characteristics
   a. Number of Floors: 42
   b. Building Height: 504 ft   154 m
   c. Per Floor Area: 40,000 ft^2 3,716 m^2

II. Passenger Elevator Requirements
   a. Low Rise Elevators:
      i. Number of Elevators: 8
      ii. Elevator Weight: 3,500 lbs   1,588 kg
      iii. Travel Speed: 700 ft/min   3.6 m·s^{-1}
      iv. Number of Floors Served: 12
      v. Floors Serviced by Elevator: 1-12
      vi. Number of Entrances (8 elevators for 14 floors): 96
   b. Mid Low Rise Elevators:
      i. Number of Elevators: 8
      ii. Elevator Weight: 3,500 lbs   1,588 kg
      iii. Travel Speed: 1,000 ft/min   3.6 m·s^{-1}
      iv. Number of Floors Served: 11
      v. Floors Serviced by Elevator: 1, 13-22
      vi. Number of Entrances (8 elevators for 15 floors): 88
   c. Mid High Rise Elevators:
      i. Number of Elevators: 8
      ii. Elevator Weight: 3,500 lbs   1,588 kg
      iii. Travel Speed: 1,200 ft/min   3.6 m·s^{-1}
      iv. Number of Floors Served: 11
      v. Floors Serviced by Elevator: 1, 23-32
      vi. Number of Entrances (8 elevators for 15 floors): 88
   d. High Rise Elevators:
      i. Number of Elevators: 8
      ii. Elevator Weight: 3,500 lbs   1,588 kg
      iii. Travel Speed: 1,400 ft/min   3.6 m·s^{-1}
      iv. Number of Floors Served: 11
      v. Floors Serviced by Elevator: 1, 33-42
      vi. Number of Entrances (8 elevators for 15 floors): 88

III. Water Protection Costs

   a. Installation of a linear drainage channel in front of each elevator hoistway and associated elevator landing:

i. Cost per hoistway/landing		$500
ii. Number of hoistways/landings (8 low rise elevators traveling 12 floors, 8 mid low elevators traveling 22 floors, 8 mid high rise elevators traveling 32 floors, and 8 high rise elevators traveling 42 floors)	x	864
iii. Total		$432,000

   b. Installation of interlocking elements on elevator frames and door panels to provide seal for the elevator

i. Cost per entrance		$250
ii. Number of entrances (8 low rise elevators serving 12 floors, 8 mid low elevators serving 11 floors, 8 mid high rise elevators serving 11 floors, and 8 high rise elevators serving 11 floors)	x	360
iii. Total		$90,000

   c. Installation of weatherproof door interlocks and wiring

i. Cost per entrance		$500
ii. Number of entrances (8 low rise elevators serving 12 floors, 8 mid low elevators serving 11 floors, 8 mid high rise elevators serving 11 floors, and 8 high rise elevators serving 11 floors)	x	360
iii. Total		$180,000

   d. Total water protection costs   **$702,000**

IV. Signage, Lobby Status Indicator, and Two-Way Communication System

a. Cost per elevator lobby		$4,500
b. Number of lobbies (above ground floor)	x	41
c. Total		**$184,500**

V. Protection of Wiring or Cables

a. Cost per elevator		$300
b. Number of elevators	x	32
c. Total		**$9,600**

VI. Lobby Enclosure Costs
a. Installation of two hollow-core steel fire doors and frame system with vision panels and automatic closing devices:

i. Cost per set		$5,250
ii. Sets per lobby		4
iii. Number of lobbies (above ground floor)	x	41
iv. Total		$861,000

b. Sealing of interstitial spaces:

i. Cost per passenger elevator lobby		$4,000 ($25 per linear foot)
ii. Number of lobbies (above ground floor)	x	41
iii. Total		$164,000

c. Total lobby enclosure costs   **$1,025,000**

VII. Maintenance Costs

a. Cost per elevator (above and beyond a passenger elevator)		$900
b. Number of passenger elevators	x	32
c. Total		**$28,800**

VIII. Initial Capital Costs of Converting passenger Elevators to Occupant Evacuation Elevators

a. Water protection	$702,000
b. Signage, lobby status indicator, and two-way communication	$184,500
c. Protection of wiring/cables	$9,600
d. Lobby enclosure	$1,025,000
e. Total	**$1,921,100**

IX. Annually Recurring Costs of Converting passenger Elevators to Occupant Evacuation Elevat

a. Maintenance	$28,800
b. Total	**$28,800**

Summary of Cost Estimates for Converting Passenger Elevators to Occupant Evacuation Elevators: Building 5

I. Characteristics
  a. Number of Floors: 75
  b. Building Height: 900 ft   274 m
  c. Per Floor Area: 45,000 ft^2 4,181 m^2

II. Passenger Elevator Requirements
  a. Low-Zone Low Rise Elevators:
    i. Number of Elevators   8
    ii. Elevator Weight: 3,500 lbs   1,588 kg
    iii. Travel Speed   500 ft/min   3.6 m·s^{-1}
    iv. Number of Floors Served   12
    v. Floors Serviced by Elevator   1-12
    vi. Number of Entrances (8 elevators for 12 floors)   96
  b. Low-Zone Mid Low Rise Elevators:
    i. Number of Elevators   8
    ii. Elevator Weight: 3,500 lbs   1,588 kg
    iii. Travel Speed   700 ft/min   3.6 m·s^{-1}
    iv. Number of Floors Served   11
    v. Floors Serviced by Elevator   1, 13-22
    vi. Number of Entrances (8 elevators for 11 floors)   88
  c. Low-Zone Mid High Rise Elevators:
    i. Number of Elevators   7
    ii. Elevator Weight: 3,500 lbs   1,588 kg
    iii. Travel Speed   1,000 ft/min   3.6 m·s^{-1}
    iv. Number of Floors Served   9
    v. Floors Serviced by Elevator   1, 23-30
    vi. Number of Entrances (7 elevators for 9 floors)   63
  d. Low-Zone High Rise Elevators:
    i. Number of Elevators   7
    ii. Elevator Weight: 3,500 lbs   1,588 kg
    iii. Travel Speed   1,200 ft/min   3.6 m·s^{-1}
    iv. Number of Floors Served   9
    v. Floors Serviced by Elevator   1, 31-38
    vi. Number of Entrances (7 elevators for 9 floors)   63

e. Shuttle Elevators:
   i. Number of Elevators   14
   ii. Elevator Weight: 10,000 lbs   4,536 kg
   iii. Travel Speed   1,200 ft/min   3.6 m·s^{-1}
   iv. Number of Floors Served   2
   v. Floors Serviced by Elevator   1, 39
   vi. Number of Entrances (7 elevators for 9 floors)   28

f. High-Zone Low Rise Elevators:
   i. Number of Elevators   8
   ii. Elevator Weight: 3,500 lbs   1,588 kg
   iii. Travel Speed   500 ft/min   3.6 m·s^{-1}
   iv. Number of Floors Served   12
   v. Floors Serviced by Elevator   39-50
   vi. Number of Entrances (8 elevators for 12 floors)   96

g. High-Zone Mid Low Rise Elevators:
   i. Number of Elevators   7
   ii. Elevator Weight: 3,500 lbs   1,588 kg
   iii. Travel Speed   700 ft/min   3.6 m·s^{-1}
   iv. Number of Floors Served   10
   v. Floors Serviced by Elevator   39, 51-59
   vi. Number of Entrances (7 elevators for 10 floors)   70

h. High-Zone Mid High Rise Elevators:
   i. Number of Elevators   7
   ii. Elevator Weight: 3,500 lbs   1,588 kg
   iii. Travel Speed   700 ft/min   3.6 m·s^{-1}
   iv. Number of Floors Served   9
   v. Floors Serviced by Elevator   39, 60-67
   vi. Number of Entrances (7 elevators for 9 floors)   63

i. High-Zone High Rise Elevators:
   i. Number of Elevators   7
   ii. Elevator Weight: 3,500 lbs   1,588 kg
   iii. Travel Speed   1,200 ft/min   3.6 m·s^{-1}
   iv. Number of Floors Served   9
   v. Floors Serviced by Elevator   39, 68-75
   vi. Number of Entrances (7 elevators for 9 floors)   63

III. Water Protection Costs
a. Total number of hoistways/landings:

  i. Number of hoistways/landings Low-Zone (8 low rise elevators traveling 12 floors, 8 mid low elevators traveling 22 floors, 7 mid high rise elevators traveling 30 floors, and 7 high rise elevators traveling 38 floors)   748

  ii. Number of hoistways/landings Shuttle (14 shuttle elevators traveling 39 floors)   546

  iii. Number of hoistways/landings High-Zone (8 low rise elevators traveling 12 floors, 7 mid low elevators traveling 21 floors, 7 mid high rise elevators traveling 29 floors, and 7 high rise elevators traveling 37 floors) +   705

  iv. Total hoistways/landings   1,999

b. Installation of a linear drainage channel in front of each elevator hoistway and associated elevator landing:
  i. Cost per hoistway/landing $500
  ii. Number of hoistways/landings   x   1,999
  iii. Total $999,500

c. Installation of interlocking elements on elevator frames and door panels to provide seal for the elevator
  i. Cost per entrance $250
  ii. Number of entrances (two zones and shuttle elevators)   x   630
  iii. Total $157,500

d. Installation of weatherproof door interlocks and wiring devices on the hoistway doors:
  i. Cost per entrance $500
  ii. Number of entrances (two zones and shuttle elevators)   x   630
  iii. Total $315,000

e. Total water protection costs   **$1,472,000**

IV. Signage, Lobby Status Indicator, and Two-Way Communication System
  a. Cost per elevator lobby $4,500
  b. Number of lobbies (above ground floor, including 2 separate lobbies for the shuttle elevators)   x   76
  c. Total   **$342,000**

V. Protection of Wiring or Cables
   a. Cost per elevator $300
   b. Number of elevators   x   73
   c. Total   **$21,900**

VI. Lobby Enclosure Costs
a. Installation of two hollow-core steel fire doors and frame system with vision panels and automatic closing devices:
   i. Cost per set $5,250
   ii. Sets per lobby   4
   iii. Number of lobbies (above ground floor)   x   76
   iv. Total $1,596,000

b. Sealing of interstitial spaces:
   i. Cost per passenger elevator lobby $4,000 ($25 per linear foot)
   ii. Number of lobbies (above ground floor)   x   76
   iii. Total $304,000

c. Total lobby enclosure costs   **$1,900,000**

VII. Sky Lobby Increment
a. Installation of two hollow-core steel fire doors and frame system with vision panels and automatic closing devices:
   i. Cost per set $5,250
   ii. Sets per lobby   x   4
   iii. Total $21,000

b. Sealing of 160 linear feet of interstitial space:
   i. Cost per linear foot $25
   ii. Total linear feet   x   160
   iii. Total $4,000

c. Total cost of protecting sky lobbies   **$25,000**

VIII. Maintenance Costs
   a. Cost per elevator (above and beyond a passenger elevator) $900
   b. Number of passenger elevators   x   73
   c. Total   **$65,700**

IX. Loss of Rental Income Due to Sky Lobby Increment
   a. Incremental loss $36.92 /ft^2 $121.13 /m^2
   b. Additional space needed for sky lobbies   x   4,000 ft^2 372 m^2
   c. Total   **$147,680**

X. Sky Lobby (Floor 39) Dimensions

a.	Gross area	45,000 gsf	4,181 m²
b.	Ratio of gross area to net area	0.85 nsf/gsf	0.08 nsm/gsm
c.	Average occupancy density	160 nsf/person	15 nsm/person
d.	Population per floor (gross area multiplied by the ratio of gross area to net area. The product is then divided by the average occupancy density.)	239	
e.	Occupant load (the number of people served by the shuttle elevators is calculated by taking the 37 floors that people are being shuttled, multiplied by the population per floor)	8,845	
f.	Area required for occupant load (3ft² per person at 25 % occupant load)	6,634 ft²	616 m²
g.	Additional area required for wheelchair accessability (10 ft²/50 people)	1,769 ft²	164 m²
h.	Sky lobby area (area required for occupant load plus additional area required for wheelchair accessability)	8,403 ft²	781 m²
i.	Anticipated lost rental income space	4,000 ft²	372 m²

XI. Initial Capital Costs of Converting passenger Elevators to Occupant Evacuation Elevators

a.	Water protection	$1,472,000
b.	Signage, lobby status indicator, and two-way communication	$342,000
c.	Protection of wiring/cables	$21,900
d.	Lobby enclosure	$1,900,000
e.	Sky Lobby Increment	$25,000
f.	**Total**	**$3,760,900**

XII. Annually Recurring Costs of Converting passenger Elevators to Occupant Evacuation Elevators

a.	Maintenance	$65,700
b.	Loss of Rental Income Due to Sky Lobby Increment	147,680
c.	**Total**	**$213,380**

Appendix B Summary of Cost Estimates for Converting Service Elevators to Fire Service Access Elevators

This appendix contains the cost estimates for converting service elevators to fire service access elevators in buildings 2 through 5. These costs include water protection costs, signage, lobby status indicators, two-way communication systems, protection of wiring, lobby enclosures, and annual maintenance costs. The table for each building is broken up into sections. Section I describes characteristics of the building, section II describes the elevator requirements for the building, and the final two sections include the total initial capital costs and the total annually recurring costs. The remaining sections describe individual costs. When applicable, metric units have been provided in a column on the right side of the page.

Building 5 has two zones of elevators. The low-zone elevators service floors 1 through 38 while the non-zone elevators service all floors. This is different than the passenger elevators, where each zone has elevators that service different floors. Buildings 2, 3, and 4 are the only buildings to have service elevators that service all floors.

Some cost items are present on every floor that an elevator travels through while others are only present on floors that an elevator stops. Hoistways/landings are present on every floor that an elevator travels through while entrances and lobbies occur only on the floors that an elevator stops. Water protection costs include the installation of a linear drainage channel in front of each elevator hoistway and associated elevator landing. These occur at every floor an elevator travels through. Water protection costs also include installation of interlocking elements on elevator frames and door panels to provide seal for the elevator entrance and installation of door interlocks and wiring devices on the hoistway doors. These occur at each floor that an elevator stops.

Summary of Cost Estimates for Converting Service Elevators to Fire Service Access Elevators: Building 2

I. Characteristics
    a. Number of Floors: 13
    b. Building Height: 156 ft   48 m
    c. Per Floor Area: 25,000 ft^2 2,323 m^2

II. Service Elevator Requirements
    a. Number of Elevators: 2
    b. Elevator Weight: 5,000 lbs   2,268 kg
    c. Travel Speed: 700 ft/min   3.6 m·s^{-1}
    d. Number of Floors Served: 13
    e. Entrances (2 elevators for 13 floors)   26

III. Water Protection Costs
    a. Installation of a linear drainage channel in front of each elevator hoistway and associated elevator landing:
      i. Cost per hoistway/landing $500
      ii. Number of hoistways/landings (2 elevators traveling 13 floors)   x   26
      iii. Total $13,000
    b. Installation of interlocking elements on elevator frames and door panels to provide seal for the elevator entrances:
      i. Cost per entrance $250
      ii. Number of entrances (2 elevators serving 13 floors)   x   26
      iii. Total $6,500
    c. Installation of weatherproof door interlocks and wiring devices on the hoistway doors:
      i. Cost per entrance $500
      ii. Number of entrances (2 elevators serving 13 floors)   x   26
      iii. Total $13,000

    d. Total water protection costs   **$32,500**

IV. Signage, Lobby Status Indicator, and Two-Way Communication System
    a. Cost per elevator lobby $4,500
    b. Number of lobbies (above ground floor)   x   12
    c. Total   **$54,000**

V. Protection of Wiring or Cables
    a. Cost per elevator $300
    b. Number of elevators   x   2
    c. Total   **$600**

VI. Lobby Enclosure Costs

a. Installation of two hollow-core steel fire doors and frame system with vision panels and automatic closing devices:
   i. Cost per set $5,250
   ii. Set per lobby   1
   iii. Number of lobbies (above ground floor)   x   12
   iv. Total $63,000

b. Sealing of interstitial spaces:
   i. Cost per service elevator lobby $1,000 ($25 per linear foot)
   ii. Number of lobbies (above ground floor)   x   12
   iii. Total $12,000

c. Total lobby enclosure costs   **$75,000**

VII. Maintenance Costs

a. Cost per elevator (above and beyond a passenger elevator) $900
b. Number of service elevators   x   2
c. Total   **$1,800**

VIII. Initial Capital Costs of Converting Service Elevators to Fire Service Access Elevators

a. Water protection $32,500
b. Signage, lobby status indicator, and two-way communication $54,000
c. Protection of wiring/cables $600
d. Lobby enclosure $75,000
e. Total   **$162,100**

IX. Annually Recurring Costs of Converting Service Elevators to Fire Service Access Elevators

a. Maintenance $1,800
b. Total   **$1,800**

Summary of Cost Estimates for Converting Service Elevators to Fire Service Access Elevators: Building 3

I. Characteristics
    a. Number of Floors: 28
    b. Building Height: 336 ft   102 m
    c. Per Floor Area: 30,000 ft^2 2,787 m^2

II. Service Elevator Requirements
    a. Number of Elevators   2
    b. Elevator Weight: 5,000 lbs   2,268 kg
    c. Travel Speed   700 ft/min   3.6 m·s^{-1}
    d. Number of Floors Served   28
    e. Floors Serviced by Elevator   1-28
    f. Number of Entrances (2 elevators for 28 floors)   56

III. Water Protection Costs
    a. Installation of a linear drainage channel in front of each elevator hoistway and associated elevator landing:
        i. Cost per hoistway/landing $500
        ii. Number of hoistways/landings (2 elevators traveling 28 floors)   x   56
        iii. Total $28,000
    b. Installation of interlocking elements on elevator frames and door panels to provide seal for the elevator entrances:
        i. Cost per entrance $250
        ii. Number of entrances (2 elevators serving 28 floors)   x   56
        iii. Total $14,000
    c. Installation of weatherproof door interlocks and wiring devices on the hoistway doors:
        i. Cost per entrance $500
        ii. Number of entrances (2 elevators serving 28 floors)   x   56
        iii. Total $28,000

    d. Total water protection costs   **$70,000**

IV. Signage, Lobby Status Indicator, and Two-Way Communication System
    a. Cost per elevator lobby $4,500
    b. Number of lobbies (above ground floor)   x   27
    c. Total   **$121,500**

V. Protection of Wiring or Cables
    a. Cost per elevator $300
    b. Number of elevators   x   2
    c. Total   **$600**

VI. Lobby Enclosure Costs
    a. Installation of two hollow-core steel fire doors and frame system with vision panels and automatic closing devices:
        i. Cost per set $5,250
        ii. Sets per lobby   1
        iii. Number of lobbies (above ground floor)   x   27
        iv. Total $141,750
    b. Sealing of interstitial spaces:
        i. Cost per service elevator lobby $1,000 ($25 per linear foot)
        ii. Number of lobbies (above ground floor)   x   27
        iii. Total $27,000
    c. Total lobby enclosure costs   **$168,750**

VII. Maintenance Costs
    a. Cost per elevator (above and beyond a passenger elevator) $900
    b. Number of service elevators   x   2
    c. Total   **$1,800**

VIII. Initial Capital Costs of Converting Service Elevators to Fire Service Access Elevators
    a. Water protection $70,000
    b. Signage, lobby status indicator, and two-way communication $121,500
    c. Protection of wiring/cables $600
    d. Lobby enclosure $168,750
    e. Total   **$360,850**

IX. Annually Recurring Costs of Converting Service Elevators to Fire Service Access Elevators
    a. Maintenance $1,800
    b. Total   **$1,800**

Summary of Cost Estimates for Converting Service Elevators to Fire Service Access Elevators: Building 4

I. Characteristics
    a. Number of Floors: 42
    b. Building Height: 504 ft   154 m
    c. Per Floor Area: 40,000 ft^2 3,716 m^2

II. Service Elevator Requirements
    a. Number of Elevators: 3
    b. Elevator Weight: 5,000 lbs   2,268 kg
    c. Travel Speed: 700 ft/min   3.6 m·s^{-1}
    d. Number of Floors Served: 43
    e. Floors Serviced by Elevator: Floor B, 1-42
    f. Number of Entrances (3 elevators for 43 floors): 129

III. Water Protection Costs
    a. Installation of a linear drainage channel in front of each elevator hoistway and associated elevator landing:
        i. Cost per hoistway/landing: $500
        ii. Number of hoistways/landings (3 elevators traveling 43 floors): x 129
        iii. Total: $64,500
    b. Installation of interlocking elements on elevator frames and door panels to provide seal for the elevator entrances:
        i. Cost per entrance: $250
        ii. Number of entrances (3 elevators serving 43 floors): x 129
        iii. Total: $32,250
    c. Installation of weatherproof door interlocks and wiring
        i. Cost per entrance: $500
        ii. Number of entrances (3 elevators serving 43 floors): x 129
        iii. Total: $64,500

    d. Total water protection costs: $161,250

IV. Signage, Lobby Status Indicator, and Two-Way Communication System
    a. Cost per elevator lobby: $4,500
    b. Number of lobbies (above ground floor): x 42
    c. Total: $189,000

V. Protection of Wiring or Cables

a. Cost per elevator		$300
b. Number of elevators	x	3
c. Total		**$900**

VI. Lobby Enclosure Costs

a. Installation of two hollow-core steel fire doors and frame system with vision panels and automatic closing devices:

i. Cost per set		$5,250
ii. Sets per lobby		1
iii. Number of lobbies (above ground floor)	x	42
iv. Total		$220,500

b. Sealing of interstitial spaces:

i. Cost per service elevator lobby		$1,000 ($25 per linear foot)
ii. Number of lobbies (above ground floor)	x	42
iii. Total		$42,000

c. Total lobby enclosure costs — **$262,500**

VII. Maintenance Costs

a. Cost per elevator (above and beyond a passenger elevator)		$900
b. Number of service elevators	x	3
c. Total		**$2,700**

VIII. Initial Capital Costs of Converting Service Elevators to Fire Service Access Elevators

a. Water protection	$161,250
b. Signage, lobby status indicator, and two-way communication	$189,000
c. Protection of wiring/cables	$900
d. Lobby enclosure	$262,500
e. Total	**$613,650**

IX. Annually Recurring Costs of Converting Service Elevators to Fire Service Access Elevators

a. Maintenance	$2,700
b. Total	**$2,700**

Summary of Cost Estimates for Converting Service Elevators to Fire Service Access Elevators: Building 5

I. Characteristics
    a. Number of Floors: 75
    b. Building Height: 900 ft 274 m
    c. Per Floor Area: 45,000 ft² 4,181 m²

II. Service Elevator Requirements
    a. Low-Zone Service Elevators:
        i. Number of Elevators 2
        ii. Elevator Weight: 5,000 lbs 2,268 kg
        iii. Travel Speed 500 ft/min 3.6 m·s^{-1}
        iv. Number of Floors Served 39
        v. Floors Serviced by Elevator Floor B, 1-38
        vi. Number of Entrances (2 elevators for 39 floors) 78
    b. Non-Zone Service Elevators:
        i. Number of Elevators 3
        ii. Elevator Weight: 5,000 lbs 2,268 kg
        iii. Travel Speed 700 ft/min 3.6 m·s^{-1}
        iv. Number of Floors Served 76
        v. Floors Serviced by Elevator Floor B, 1-75
        vi. Number of Entrances (3 elevators for 76 floors) 228

III. Water Protection Costs
    a. Total number of hoistways/landings:
        i. Number of hoistways/landings Low-Zone (2 elevators traveling 39 floors) 78
        ii. Number of hoistways/landings Non-Zone (3 elevators traveling 76 floors) + 228
        iv. Total hoistways/landings 306
    b. Installation of a linear drainage channel in front of each elevator hoistway and associated elevator landing:
        i. Cost per hoistway/landing $500
        ii. Number of hoistways/landings x 306
        iii. Total $153,000
    c. Installation of interlocking elements on elevator frames and door panels to provide seal for the elevator entrances:
        i. Cost per entrance $250
        ii. Number of entrances (low-zone and non-zone) x 306
        iii. Total $76,500

   d. Installation of weatherproof door interlocks and wiring
   devices on the hoistway doors:

i. Cost per entrance		$500
ii. Number of entrances (low-zone and non-zone)	x	306
iii. Total		$153,000

   e. Total water protection costs   **$382,500**

IV. Signage, Lobby Status Indicator, and Two-Way Communication System

a. Cost per elevator lobby		$4,500
b. Number of lobbies (above ground floor)	x	113
c. Total		**$508,500**

V. Protection of Wiring or Cables

a. Cost per elevator		$300
b. Number of elevators	x	5
c. Total		**$1,500**

VI. Lobby Enclosure Costs

  a. Installation of two hollow-core steel fire doors and frame
  system with vision panels and automatic closing devices:

i. Cost per set		$5,250
ii. Sets per lobby		1
iii. Number of lobbies (above ground floor)	x	113
iv. Total		$593,250

  b. Sealing of interstitial spaces:

i. Cost per service elevator lobby		$1,000 ($25 per linear foot)
ii. Number of lobbies (above ground floor)	x	113
iii. Total		$113,000

  c. Total lobby enclosure costs   **$706,250**

VII. Maintenance Costs

a. Cost per elevator (above and beyond a passenger elevator)		$900
b. Number of service elevators	x	5
c. Total		**$4,500**

VIII. Initial Capital Costs of Converting Service Elevators to Fire Service Access Elevators

a. Water protection	$382,500
b. Signage, lobby status indicator, and two-way communication	$508,500
c. Protection of wiring/cables	$1,500
d. Lobby enclosure	$706,250
e. Total	**$1,598,750**

IX. Annually Recurring Costs of Converting Service Elevators to Fire Service Access Elevators

a. Maintenance	$4,500
b. Total	**$4,500**

www.ingramcontent.com/pod-product-compliance
Lightning Source LLC
Chambersburg PA
CBHW081726170526
45167CB00009B/3712